# Optimizing Smart and Sustainable Agriculture for Sustainability

This reference text addresses the importance of smart crop management for increasing yield and presents a framework for smart monitoring and regulation of crop observation. Further, it comprehensively covers important topics such as spatial decision support systems for precision farming, swarm intelligence in the optimal management of aquaculture farms, and intelligent harvesting algorithms for improving productivity.

This book:

- Presents meta-heuristic algorithms for optimization, economic crop planning, and use of effective water resource management.
- Discusses spatial decision support systems for crop productivity management, watershed management, and precision farming.
- Illustrates swarm intelligence-based optimization techniques, data mining, and machine learning methods for aquaculture operations.
- Highlights artificial intelligence and machine learning-based harvesting algorithms for improving productivity.
- Explains the use of green Internet of Things security solutions for agriculture, plant condition management, and greenhouse simulation.

It is primarily written for graduate students and academic researchers in the fields of electrical engineering, electronics and communication engineering, computer science and engineering, agricultural science, and information technology.

**Future Generation Information Systems**
**Series Editor- Bharat Bhushan**

*With the evolution of future generation computing systems, it becomes necessary to occasionally take stock, analyze the development of its core theoretical ideas, and adapt to radical innovations. This series will provide a platform to reflect the theoretical progress and forge emerging theoretical avenues for the future generation information systems. The theoretical progress in the Information Systems field (IS) and the development of associated next generation theories is the need of the hour. This is because Information Technology (IT) has become increasingly infused, interconnected and intelligent in almost all context.*

**Industry 6.0: Technology, Practices, Challenges, and Applications**
*Kishor Kumar Reddy C, Srinath Doss, Lavanya Pamulaparty, Kari J Lippert and Ruchi Doshi*

**Convergence of Deep Learning and Artificial Intelligence in Internet of Things**
*Ajay Rana, Arun Rana, Sachin Dhawan, Sharad Sharma and Ahmed A. Elngar*

***Next Generation Communication Networks for Industrial Internet of Things Systems***
*Sundresan Perumal, Mujahid Tabassum, Moolchand Sharma and Saju Mohanan*

***Artificial Intelligence and Optimizing Techniques for Smart Information System Generations***
Edited by: Aleem Ali, Rajdeep Chakraborty and Nawaf R. Alharbe

***Artificial Intelligence for Healthcare Applications***
Edited by: Anoop V.S, Suhasini Verma and Usharani Hareesh Govindarajan

***Optimizing Smart and Sustainable Agriculture for Sustainability***
Edited by: Biswaranjan Acharya, Ankita Bansal, Abha Jain, Rachna Jain and Joel J. P. C. Rodrigues

# Optimizing Smart and Sustainable Agriculture for Sustainability

*Edited by*
Biswaranjan Acharya
Ankita Bansal
Abha Jain
Rachna Jain
Joel J. P. C. Rodrigues

CRC Press
Taylor & Francis Group
Boca Raton London New York

CRC Press is an imprint of the
Taylor & Francis Group, an **informa** business

Designed cover image: shutterstock

First edition published 2025
by CRC Press
2385 NW Executive Center Drive, Suite 320, Boca Raton FL 33431

and by CRC Press
4 Park Square, Milton Park, Abingdon, Oxon, OX14 4RN

*CRC Press is an imprint of Taylor & Francis Group, LLC*



ISBN: 978-1-032-65794-3 (hbk)
ISBN: 978-1-041-04482-6 (pbk)
ISBN: 978-1-003-62839-2 (ebk)

DOI: 10.1201/9781003628392

Typeset in Sabon
by SPi Technologies India Pvt Ltd (Straive)

# Contents

# About the Editors

**Biswaranjan Acharya** earned his M.C.A. degree from IGNOU, New Delhi, India, in 2009; an M.Tech. degree in Computer Science and Engineering from Biju Patnaik University of Technology (BPUT), Rourkela, Odisha, India, in 2012; and a Ph.D. in Computer Science from Veer Surendra Sai University of Technology (VSSUT), Burla, Odisha, India, in 2024. He is an assistant professor in the Department of Computer Engineering-AI and BDA at Marwadi University, Gujarat, India. He received a research fellowship at INTI International University from December 15, 2023, to December 31, 2025. He has more than 10 years of academic experience at reputed institutions such as Ravenshaw University and has also worked in the software development industry. He has coauthored more than 70 research articles in internationally reputed journals and serves as a reviewer for several peer-reviewed journals. Additionally, he holds more than 50 patents. His research interests include multiprocessor scheduling, data analytics, computer vision, machine learning, and the Internet of Things (IoT). He is currently serving as a secondary IEEE Computer Society representative to the IEEE Nanotechnology Council (NTC) Administrative Committee and as an observer of the IEEE P2851 Standard for Functional Safety Data Format. He is also associated with various educational and research societies, including IACSIT, CSI, IAENG, and ISC. Google Scholar Profile: Biswaranjan Acharya (ବିଶ୍ୱରଂଜନ ଆଚାର୍ଯ୍ୟ) SMIEEE—Google Scholar.

**Ankita Bansal** is an assistant professor in the Department of Information Technology in Netaji Subhas University of Technology. She has teaching experience of more than 10 years. She was awarded a doctoral degree in Computer Science from Delhi Technological University (DTU, formerly DCE). From the same university, she earned her master's degree in Computer Technology and Applications and secured rank 1 in the same. She has published more than 40 papers in reputed international journals and conferences. She has several edited books to her credit. Her research interests are computational intelligence, software testing, software quality, software metrics, developing quality prediction models. Google Scholar: ankita Jain bansal—Google Scholar.

**Abha Jain** is an assistant professor in the Department of Computer Science, Dyal Singh College, University of Delhi. Prior to joining the college, she worked as full-time research scholar and received a doctoral research fellowship from Delhi Technological University (formerly Delhi College of Engineering). She earned her master's and doctorate degrees in Software Engineering from Delhi Technological University. Her research interests are data mining, software quality, and statistical and machine learning models. She has published papers in international journals and conferences.

**Rachna Jain** is an associate professor at Bhagwan Parshuram Institute of Technology (GGSIPU) since August 2021. She worked as an assistant professor (Computer Science Department) at Bharati Vidyapeeth's College of Engineering (GGSIPU) from August 2007 to August 2021.She completed her Ph.D. in Computer Science from Banasthali Vidyapith in 2017. She earned her M.E. degree in 2011 from Delhi College of Engineering (Delhi University) with specialization in Computer Technology and Applications. She completed her B.Tech (Computer Science) in 2006 from N.C. College of Engineering, Kurukshetra University. Her current research interests are cloud computing, fuzzy logic, network and information security, swarm intelligence, big data and IoT, deep learning, and machine learning. She has contributed with more than 25 book chapters in various books. She has also served as session chair in various international conferences. She was CO-PI of DST Project titled "Design an autonomous intelligent drone for city surveillance". She has a total of 18 years of academic/research experience with more than 100 publications in various national, international conferences, and international journals (Scopus/ISI/SCI) of high repute. Google Scholar link: Rachna Jain—Google Scholar.

**Joel J. P. C. Rodrigues** is with the Federal University of Piauí, Brazil, and Leader of the Center for Intelligence at Fecomércio/CE, Brazil. Prof. Rodrigues is an Highly Cited Researcher (Clarivate), N. 1 of the top scientists in computer science in Brazil (Research.com), the leader of the Next Generation Networks and Applications (NetGNA) research group (CNPq), Member Representative of the IEEE Communications Society on the IEEE Biometrics Council, and the President of the scientific council at ParkUrbis – Covilhã Science and Technology Park. He was Director for Conference Development - IEEE ComSoc Board of Governors, an IEEE Distinguished Lecturer, Technical Activities Committee Chair of the IEEE ComSoc Latin America Region Board, a Past-Chair of the IEEE ComSoc Technical Committee (TC) on eHealth and the TC on Communications Software, a Steering Committee member of the IEEE Life Sciences Technical Community and Publications co-chair. He is the editor-in-chief of the *International Journal of E-Health and Medical Communications* and editorial board member of several high-reputed journals (mainly, from IEEE). He has been general chair and TPC Chair of many international conferences, including IEEE ICC, IEEE GLOBECOM, IEEE HEALTHCOM, and IEEE

LatinCom. He has authored or coauthored about 1300 papers in refereed international journals and conferences, 3 books, 2 patents, and 1 ITU-T Recommendation. He had been awarded several Outstanding Leadership and Outstanding Service Awards by IEEE Communications Society and several best papers awards. Prof. Rodrigues is a member of the Internet Society, a senior member ACM, and Fellow of AAIA and IEEE.

# Contributors

**Biswaranjan Acharya**
Department of Computer Engineering—AI & BDA
Marwadi University
Rajkot, India

**Kakoli Banerjee**
Department of Computer Science and Engineering
JSS Academy of Technical Education
Noida, India

**Dev Batra**
JSS Academy of Technical Education Noida
Noida, India

**Sunil Kumar Chawla**
Department of Computer Science and Engineering
Chitkara University Institute of Engineering and Technology
Chitkara University
Punjab, India

**P. Sathvik Chowdari**
Sree Vidyanikethan Engineering College (Autonomous)
India

**Shailendra S. Gaur**
BPIT, GGSIPU
New Delhi, India

**Vimal Gaur**
CSE Department
Maharaja Surajmal Institute of Technology
New Delhi, India

**Vassilis C. Gerogiannis**
Department of Digital Systems
University of Thessaly
Larissa, Greece

**Debasis Gountia**
School of Computer Sciences
Odisha University of Technology and Research
Bhubaneshwar, India

**Shweta Gupta**
Department of Computer Science & Engineering
Bhagwan Parshuram Institute of Technology
Rohini, India

**Diksha Jain**
Department of Electronics and Communication Engineering
Bharati Vidyapeeth's College of Engineering
New Delhi, India

**Rachna Jain**
Department of Information Technology
Bhagwan Parshuram Institute of Technology
New Delhi, India

**Shivani Jha**
Division of Plant Pathology
Indian Agricultural Research Institute
New Delhi, India

**Divyanshu Joshi**
JSS Academy of Technical Education Noida
Noida, India

**Namburi Karunakar Reddy**
Department of Plant Pathology
University of Agricultural Sciences, GKVK
Bangalore, India

**Jaspreet Kaur**
JSS Academy of Technical Education Noida
Noida, India

**Ushveen Kaur**
Department of Computer Science
Sri Guru Gobind Singh College of Commerce
University of Delhi
New Delhi, India

**Ajay Kumar**
Department of Computer Science and Engineering
JSS Academy of Technical Education
Noida, India

**Pradeep Kumar**
Department of Computer Science and Engineering
JSS Academy of Technical Education
Noida, India

**B. Narendra Kumar Rao**
Mohan Babu University
Tirupati, India

**Mallela Nikhil Kumar Reddy**
Sree Vidyanikethan Engineering College (Autonomous)
Tirupati, India

**Sanjeev Kumar**
MAIT, GGSIPU, CH Bhim Singh Nambardar Marg
Delhi, India

**Jyotirmay Mahapatra**
Department of Farm Machinery and Power Engineering
College of Agricultural Engineering and Technology
Odisha University of Agriculture and Technology
Bhubaneswar, India

**Divyanshi Mishra**
Department of Agricultural Economics
College of Agriculture
Acharya Narendra Deva University of Agriculture and Technology
Ayodhya (U.P.), India

**Harshit Mishra**
Department of Agricultural Economics
College of Agriculture
Acharya Narendra Deva University of Agriculture and Technology
Ayodhya (U.P.), India

**Manas Mohanty**
Teamware Solution
Chennai, India

**Jolly Parikh**
Department of Computer Science and Engineering
Bharati Vidyapeeth's College of Engineering
New Delhi, India

**Naide Prathap Reddy**
Krishi Vigyan Kendra
Tuniki, Kowdipally, Medak-Telangana, India

**Pooja Purushotham**
Department of Plant Pathology
University of Agricultural Sciences, GKVK
Bangalore, India

**Shipra Raheja**
IT Department
Guru Tech Bahadur Institute of Technology
New Delhi, India

**Gaurav Y. Rakhonde**
Department of Plant Pathology
University of Agricultural Sciences, GKVK
Bangalore, India

**Prabhakara S.**
IIBF
Mumbai, India

**Amaresh Sahu**
Department of Computer Science and Application
Ajay Binay Institute of Technology
Cuttack, India

**Partha Sarathi Sahoo**
Department of Computer Science and Application
Ajay Binay Institute of Technology
Cuttack, India

**Ranjit Singh Sarban Singh**
Department of Computing and Information System
School of Engineering and Technology
Sunway University
Malaysia

**Mrutyunjay Senapati**
School of Computer Sciences
Odisha University of Technology and Research
Bhubaneshwar, India

**Chaitanya Sharma**
JSS Academy of Technical Education Noida
Noida, India

**Nidhi Sharma**
Department of Applied Sciences
VIPS-Technical Campus
Delhi, India

**Murthy Chetty Muni Sreenivas**
Sree Vidyanikethan Engineering College (Autonomous)
Tirupati, India

**Somireddy Srinivasa Reddy**
Professor Jayashankar Telangana State Agricultural University
Rajendranagar Mandal
Hyderabad, India

**Bhawna Suri**
Department of Computer Science & Engineering
Bhagwan Parshuram Institute of Technology
Rohini, India

**Shweta Taneja**
Department of Computer Science & Engineering
Bhagwan Parshuram Institute of Technology
Rohini, India

**Chirag Tyagi**
JSS Academy of Technical Education Noida
Noida, India

**Nidyanand Yuvaraj Singh**
Sree Vidyanikethan Engineering College (Autonomous)
Tirupati, India

# Preface

Agriculture plays a very pivotal role in the global economy. With the advent of precision farming, Internet of Things (IoT) devices are being utilized to collect data related to soil nutrients, water, and fertilizer requirements. This data is used for analysing and increasing crop growth. In addition to IoT devices, autonomous and semi-autonomous devices such as unmanned aerial vehicles are used for identifying various plant diseases which adversely affect the crop growth. Satellite images are also used in precision agriculture for monitoring the field and identifying the diseases in the plants. Autonomous tractors or robots are also used for the purpose of irrigation to increase the yield.

The goal of this book is to provide a forum advancing the state-of-the-art of machine learning and meta-heuristic design based on the important real-world applications in the field of agriculture. The areas focused on in this book are prediction of soil parameters such as organic carbon and moisture content, crop yield prediction, disease and weed detection in crops, and species detection. Precision farming is reviewed for the classification of a different set of crop images in order to monitor the crop quality and yield assessment. These are the 13 chapters included in the book.

Chapter 1 describes livestock farming that significantly impacts the global economy and food security, accounting for 40% of agricultural output. The adoption of smart technologies like IoT and machine learning has enhanced productivity, disease prevention, and cost efficiency, though challenges such as data privacy and technical barriers remain. Future trends focus on blockchain, AI, and sustainability for continued sector improvement. Chapter 2 shows that the increasing global food demand and changing land-to-labour ratio require better crop management and machinery. Technological advancements in sensors, drones, satellites, and AI/ML have enabled real-time data collection and improved decision-making in farming. These technologies help predict and detect issues like diseases, water stress, and nutrient deficiency, optimizing inputs and operations for enhanced efficiency and environmental control.

Chapter 3 suggests that recent automatic disease detection systems, especially those using deep learning (DL) like residual neural networks (ResNet), are more efficient and cost-effective than manual methods. ResNet addresses challenges like vanishing gradients, making it ideal for training deep networks and improving plant disease diagnosis. This research uses image processing and convolutional neural networks (CNN) to detect leaf diseases early, helping to reduce crop losses and improve yields. Chapter 4 presents crop yield prediction (CYP) using machine learning that can help address these issues, but model uncertainty due to complex data integration affects accuracy. This chapter proposes a Bi-LSTM deep learning model that improves CYP accuracy to 99.01%, outperforming other models like LSTM, RNN, and SVM, which have lower accuracies of 97.99%, 97.80%, and 97.49%, respectively.

Chapter 5 introduces obesity-related health issues; there's a growing emphasis on healthier eating and calorie monitoring to prevent conditions like high blood pressure, diabetes, and heart problems. This proposal suggests using deep learning technology to calculate calories from user-captured food images, employing algorithms like CNN for accurate food classification and analysis. This system helps individuals make informed dietary choices by increasing awareness of their daily calorie intake and the nutritional content of their food. Chapter 6 emphasizes smart crop management using technology like sensors and data analytics, which help optimize irrigation, fertilization, and pest control, improving yields and reducing waste. Addressing challenges like pest control and climate change is essential to enhance productivity and ensure food security, with precision agriculture being key to meeting global food demands sustainably. Chapter 7 analyses issues leading to the need for Green Revolution 2.0, emphasizing the role of good agricultural practices (GAP), sustainability, AI, and IoT in enhancing productivity and soil quality. It highlights the global need for sustainable food grain production and the importance of extending this effort to all natural resources.

Chapter 8 develops a prototype for an Edge AI device to monitor and regulate greenhouse conditions, ensuring optimal plant growth. This approach integrates modern technology with greenhouse farming, advancing the concept of smart greenhouses. Chapter 9 introduces agricapital, which provides a centralized platform that connects farmers directly with buyers. Additionally, machine learning is transforming farming practices by optimizing productivity and resource use through data-driven insights and precision farming technologies like drones and algorithms. Chapter 10 integrates data-driven operations and advanced technologies to optimize farming, enhance sustainability, and minimize environmental impact. This chapter explores the concept of smart and sustainable agriculture, reviewing literature and case studies on its implementation.

The IoT connects devices through electronics, sensors, and software, enhancing efficiency across sectors like healthcare, smart cities, and agriculture, with a projected economic impact of up to USD 6.2 trillion by 2025. Chapter 11 provides insights into the green IoT (GIoT) concerns about energy consumption and $CO_2$ emissions by improving energy efficiency and reducing environmental impact. This chapter examines GIoT's application in smart agricultural monitoring, focusing on both software and hardware techniques to optimize energy use and promote sustainable practices in smart farming. Chapter 12 describes that accurate CYPs are crucial for optimizing resource usage, minimizing risks, and enhancing sustainability in agriculture. This research utilizes a diverse dataset from five crops (rice, ragi, gram, potato, and onion) collected in Odisha, and employs machine learning and deep learning algorithms like random forest, SVM, LSTM, and RNN. Among them, random forest achieved 99.27% accuracy, while the sigmoid activation function in neural networks reached 99.71%. These methods effectively handle heterogeneous data environments, supporting better agricultural decision-making and food security. In Chapter 13, it is shown that India's agriculture sector, vital to 58% of its population, has seen significant contributions from agriculture and allied exports, with advancements in technologies such as drones playing a transformative role. Agriculture 4.0, incorporating big data, IoT, and drones, enhances productivity and sustainability by improving crop monitoring, precision spraying, and yield forecasting. Drones offer detailed field mapping and real-time crop health monitoring, enabling better decision-making and early problem detection, which is crucial for sustainable pest and disease management.

Chapter 1

# Economic efficiency through livestock production monitoring and prediction in smart agriculture

*Harshit Mishra and Divyanshi Mishra*

## 1.1 INTRODUCTION

The livestock sector stands as a pivotal force in the global economy, constituting approximately 40% of the total value of agricultural output. Its paramount role in ensuring food security cannot be overstated, as it not only supplies essential nutrients but also generates income for millions worldwide (Su and Wang, 2021; Monteiro *et al.*, 2021). Economically, the livestock industry has undergone substantial expansion in recent decades, propelled by burgeoning demand for animal-derived products, particularly in developing nations. This growth surge is fuelled by factors such as rising incomes, urbanization, and evolving dietary preferences (O'Grady and O'Hare, 2017). In 2020, global meat production soared to an unprecedented 335 million tonnes, with poultry commanding the largest share at 46%, trailed by pork at 36%, and beef at 16% (Tedeschi *et al.*, 2021; Mishra and Sharma, 2023).

The sector has grappled with inflationary pressures in recent years, with escalating feed costs, labour expenditures, and transportation expenses contributing to an upswing in prices for animal products. The COVID-19 pandemic added another layer of complexity, disrupting supply chains and intensifying price volatility (Gupta and Nahar, 2023; Said Mohamed *et al.*, 2021). In 2022, the Food and Agriculture Organization of the United Nations (FAO) recorded its highest-ever Meat Price Index, propelled by robust demand and supply constraints. Beyond its direct economic implications, the livestock sector exerts a profound influence on the economies of both developed and developing countries. In developed nations, it contributes significantly to employment, tax revenue, and foreign exchange earnings (Choi and Shin, 2023). Conversely, in developing countries, livestock production often serves as a primary source of income and livelihood security, particularly for rural households. The intricate interplay of economic growth, inflation, and overall economic impact underscores the sector's intricate position within the global landscape (Chatterjee *et al.*, 2021 Nishad *et al.*, 2023).

DOI: 10.1201/9781003628392-1

In India, the livestock sector plays a significant role in the economy, serving as a vital contributor to both rural livelihoods and the national gross domestic product (GDP). Currently, the sector contributes approximately 4.11% to India's GDP and constitutes around 25.6% of the total agriculture GDP. Employing about 8.8% of the Indian population, the livestock sector is a crucial source of livelihood, particularly in rural areas, where over 80 million dairy farmers find direct employment, and millions more benefit indirectly (Ramya *et al.*, 2015; Perakis *et al.*, 2020). India stands as the world's largest milk producer, contributing to 16% of global milk production, with an estimated output of 198.4 million tonnes in the fiscal year 2022–23. Additionally, India plays a significant role in the production of eggs, meat, and other livestock products (Javaid *et al.*, 2022).

The economic benefits of the livestock sector extend beyond GDP contribution, with substantial income generated for farmers, especially small and marginal ones. Beyond economic gains, the sector contributes to nutritional security by providing access to high-quality protein and essential micronutrients (Monteiro *et al.*, 2021; Ukhurebor *et al.*, 2022). Moreover, livestock products are integral to India's export earnings, adding to the country's economic strength. However, the sector is not without its challenges. It faces issues such as disease outbreaks, inadequate infrastructure, and fluctuating market prices. Despite these challenges, opportunities for growth and development abound (Vrchota *et al.*, 2022). The increasing demand for livestock products, coupled with technological advancements and government initiatives aimed at supporting the sector, presents avenues for further expansion and enhancement. As India navigates the complexities of the livestock industry, a strategic approach that addresses challenges while capitalizing on opportunities will be crucial for sustaining and maximizing the sector's contributions to the nation's economy and well-being (Abdullahi *et al.*, 2015; Jaiswal *et al.*, 2019).

As an indispensable facet of agricultural practices, livestock not only contributes substantially to the global economy but also holds paramount importance in sustaining human life. This chapter address the integral role of livestock, its economic significance, and the persistent challenges faced in contemporary livestock farming (Kakamoukas *et al.*, 2021). Recognizing livestock as a cornerstone of agriculture, the discussion unfolds by elucidating its significance within the broader context of agricultural practices. Livestock emerges as not merely a source of sustenance but as a dynamic force shaping economic landscapes worldwide. The economic significance of livestock production is examined, highlighting its intricate interplay with global markets and its profound impact on the livelihoods of millions (Neethirajan and Kemp, 2021; Wolfert *et al.*, 2017; Sharma *et al.*, 2022). Amid the agricultural panorama, challenges in livestock farming pose formidable hurdles that demand attention. From disease outbreaks to environmental pressures, these challenges necessitate innovative solutions. This chapter, however, posits that the integration of technology and the principles of smart agriculture can offer transformative solutions to long-standing predicaments (Lovarelli *et al.*, 2020; Abdalla and El-Ramady, 2022).

The narrative seamlessly transitions to an exploration of smart agriculture and its technological underpinnings. Smart agriculture is unveiled, encompassing a spectrum of innovations and concepts designed to enhance efficiency and sustainability. Technological advancements in agriculture, including the integration of the Internet of Things (IoT) and sensor technologies (as shown in Figure 1.1), become focal points in the subsequent exploration, as they converge to reshape the landscape of livestock farming (Hedley, 2015; Kumar *et al.*, 2023). A crucial element in this technological metamorphosis is the establishment of robust data collection and monitoring systems. Real-time data collection in livestock farms takes a centre stage, empowered by a diverse array of sensors and IoT devices meticulously designed for livestock monitoring. The chapter then navigates through the intricacies of data storage and management solutions, underscoring the critical role they play in harnessing the potential of the amassed data (Alonso *et al.*, 2020). As the discussion evolves, the focus pivots towards the pivotal role of data analytics in livestock farming. The importance of data analytics is expounded upon, setting the stage for the exploration of machine learning and artificial intelligence (AI) in predicting the trajectory of livestock health and productivity. Predictive models, emerging as powerful tools, offer a glimpse into the future of precision livestock farming (Pathan *et al.*, 2020; Madushanki *et al.*, 2019).

The ensuing sections illuminate the tangible benefits arising from the integration of livestock production monitoring and prediction systems. Improved livestock health and well-being, increased productivity and efficiency, and cost reduction coupled with resource optimization emerge as tangible outcomes (Tiwari *et al.*, 2023). Yet, amid the promises of technological advancement, it candidly addresses the inherent challenges and limitations, including technical hurdles, data privacy concerns, and economic barriers that necessitate thoughtful consideration (Zhang and Cao, 2019; Seelan *et al.*, 2003). The narrative unfolds further into the future, exploring emerging trends and innovations in livestock farming. The integration of blockchain and AI takes a centre stage, offering a glimpse into a future where cutting-edge technologies converge to redefine the landscape of livestock production. Sustainability and environmental considerations round off this exploration, emphasizing the need for a holistic approach that balances technological innovation with ecological stewardship (Banhazi *et al.*, 2012). In the penultimate sections, it offers pragmatic recommendations and best practices for stakeholders. Guiding principles for implementing livestock monitoring and prediction systems, coupled with strategies for training and skill development for farmers, underscore the importance of a comprehensive and inclusive approach. The indispensable role of government and industry support is emphasized as a linchpin in fostering widespread adoption and success in the integration of smart technologies into

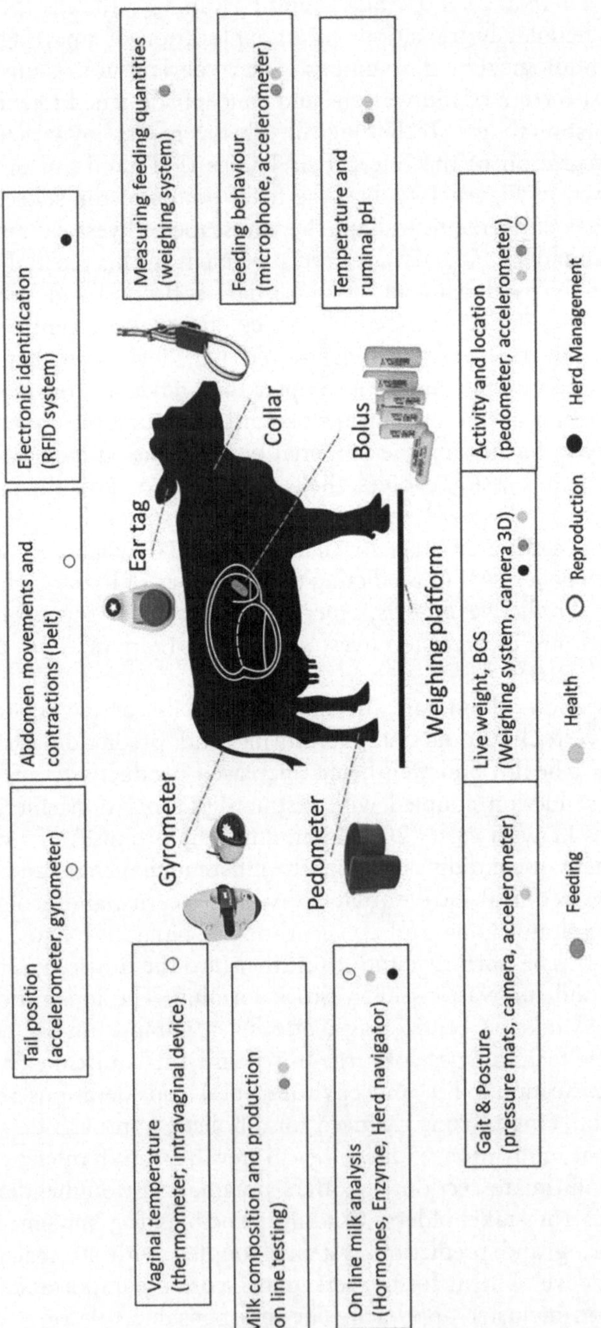


*Figure 1.1* Overview of the Internet of Things (IoT) related to dairy farming.

**Source: Kleen and Guatteo, 2023.**

livestock farming practices (Higgins *et al.*, 2019; Schellberg *et al.*, 2008; Gupta *et al.*, 2020).

## 1.2 THE ROLE OF LIVESTOCK IN AGRICULTURE

### 1.2.1 Livestock as a major component of agriculture

Livestock play a pivotal role in the overall framework of agriculture, contributing significantly to the global food supply, economic development, and sustainable agricultural practices. Figure 1.2 show the prominence of livestock in agriculture.

### 1.2.2 Economic significance of livestock production

Livestock production is not only a source of food but also a significant contributor to the economic well-being of individuals, communities, and nations. The economic significance of livestock production is multifaceted and includes the following aspects:

- **Job creation:** Livestock farming generates employment opportunities across the entire value chain, from animal husbandry and veterinary services to processing, marketing, and distribution. This, in turn, contributes to poverty reduction and rural development.
- **Income generation:** Livestock products, such as meat, milk, and leather, contribute to local and national economies by providing a source of income for farmers and entrepreneurs involved in the livestock sector. The sale of livestock and related products is a crucial revenue stream for many households.
- **Export and trade:** Livestock and their products form a significant component of international trade. Countries with well-established livestock industries often engage in the export of meat, dairy, and other products, contributing to foreign exchange earnings and economic growth.
- **Diversification of income sources:** Livestock farming allows for income diversification, reducing the vulnerability of farmers to the fluctuations in crop yields and market prices. This diversification enhances the overall economic resilience of agricultural communities.

Recognizing the central role of livestock in agriculture is essential for understanding the broader implications of livestock production on food security, economic development, and sustainable agricultural practices. Monitoring and predicting livestock production in the context of smart agriculture can further enhance these contributions by optimizing resource use, improving efficiency, and mitigating risks (Bhattacharyya *et al.*, 2020). Table 1.1 displays the challenge, impact, and potential solutions associated with livestock farming.

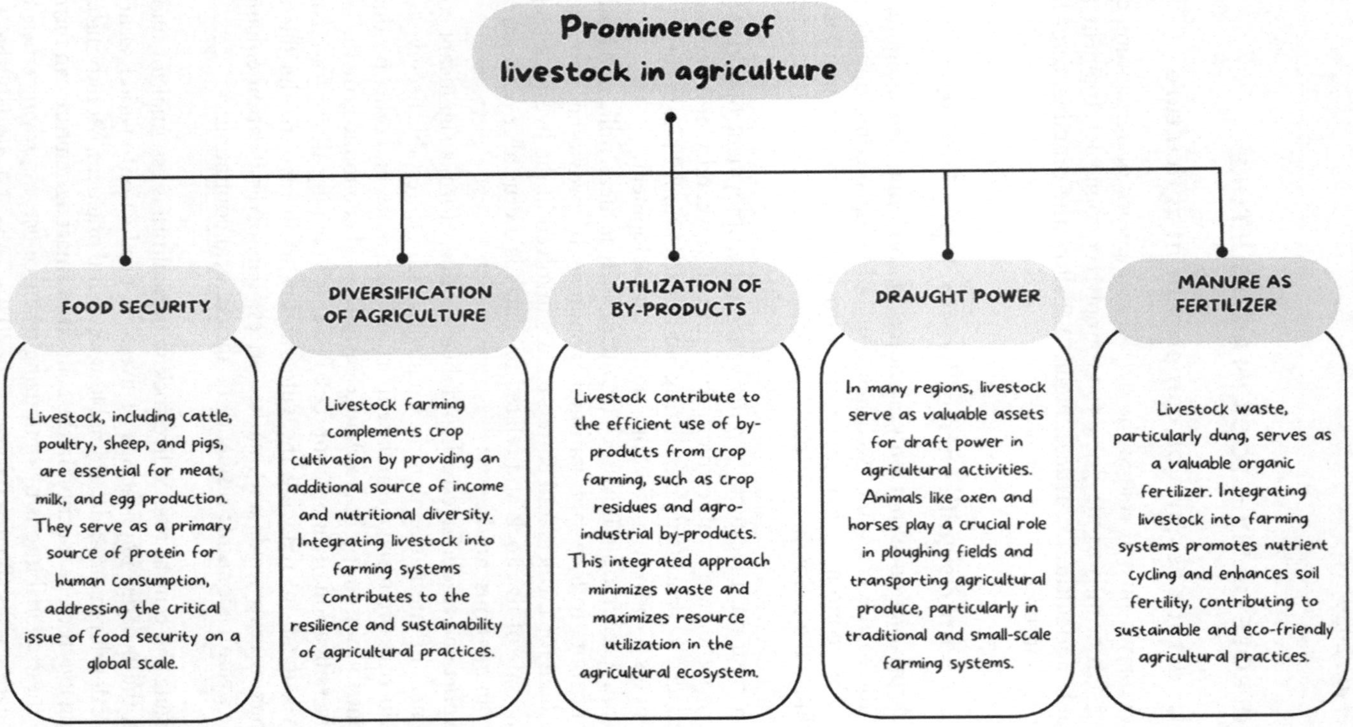


*Figure 1.2* Prominence of livestock in agriculture.

*Table 1.1* Overview of challenges in livestock farming

| *Challenge* | *Description* | *Impact* | *Potential solutions* |
|---|---|---|---|
| Feed scarcity and poor quality | Insufficient availability of nutritious feed for livestock | Reduced productivity, animal health issues, and higher production costs | Improve pasture management practices, utilize alternative feed sources, invest in feed processing technologies |
| Water scarcity and pollution | Limited access to clean water for livestock and livestock waste management issues | Waterborne diseases, environmental pollution, and reduced productivity | Implement water conservation strategies, adopt sustainable waste management practices, improve access to water sources |
| Animal diseases and zoonotic outbreaks | Prevalence of animal diseases and the risk of zoonotic transmission | Reduced productivity, economic losses, and public health threats | Enhance veterinary services, implement preventive measures, promote animal health awareness |
| Climate change and its effects | Extreme weather events, heat stress, and changes in feed availability | Impact on animal health, productivity, and feed production | Adopt climate-resilient farming practices, improve infrastructure to mitigate climate impacts, develop heat-tolerant livestock breeds |
| Market fluctuations and price volatility | Unpredictable market prices and demand fluctuations | Income instability for farmers, reduced profitability, and challenges in planning production | Improve market information systems, develop risk management strategies, promote value addition and diversification of products |
| Inadequate infrastructure and technology | Lack of access to essential infrastructure and technological advancements | Inefficient production practices, higher production costs, and challenges in quality control | Invest in infrastructure development, promote adoption of appropriate technologies, provide training and support for farmers |

## 1.3 SMART AGRICULTURE AND TECHNOLOGY INTEGRATION

Smart agriculture represents the incorporation of advanced technologies into traditional farming practices to enhance efficiency, productivity, and sustainability. In the context of livestock production, integrating smart agriculture principles can revolutionize the way farmers manage their herds. This section will delve into the key components of smart agriculture, technological advancements in the agricultural sector, and the specific application of the IoT and sensor technologies in livestock farming (Mekonnen *et al.*, 2020; Morrone *et al.*, 2022).

### 1.3.1 Concepts of smart agriculture

Smart agriculture involves the use of cutting-edge technologies to optimize various aspects of farming operations. This includes the intelligent use of data, automation, and connectivity to make informed decisions and enhance overall agricultural productivity (De Alwis *et al.*, 2022). In the context of livestock farming, smart agriculture encompasses a range of applications, from monitoring animal health and behaviour to optimizing feeding schedules and managing reproduction cycles (Singh *et al.*, 2022; Dahane *et al.*, 2022).

Concepts within smart agriculture for livestock production include:

- **Data-driven decision making:** Utilizing data analytics and farm management software to analyse real-time information about the health, behaviour, and performance of livestock. This data-driven approach enables farmers to make informed decisions for better resource utilization and management.
- **Automation and robotics:** Implementing automated systems and robotics in tasks such as feeding, milking, and waste management. This reduces the need for manual labour, increases efficiency, and minimizes human errors.
- **Precision livestock farming:** As illustrated in Figure 1.3, the application of precision agriculture principles to livestock management entails leveraging Global Positioning System (GPS), sensors, and data analytics. This approach facilitates the monitoring and optimization of various aspects such as feeding, breeding, and health management.

### 1.3.2 Technology advancements in agriculture

Technological advancements in agriculture have played a pivotal role in transforming traditional farming practices. This subsection will explore some of the key technologies that have contributed to the evolution of smart agriculture:

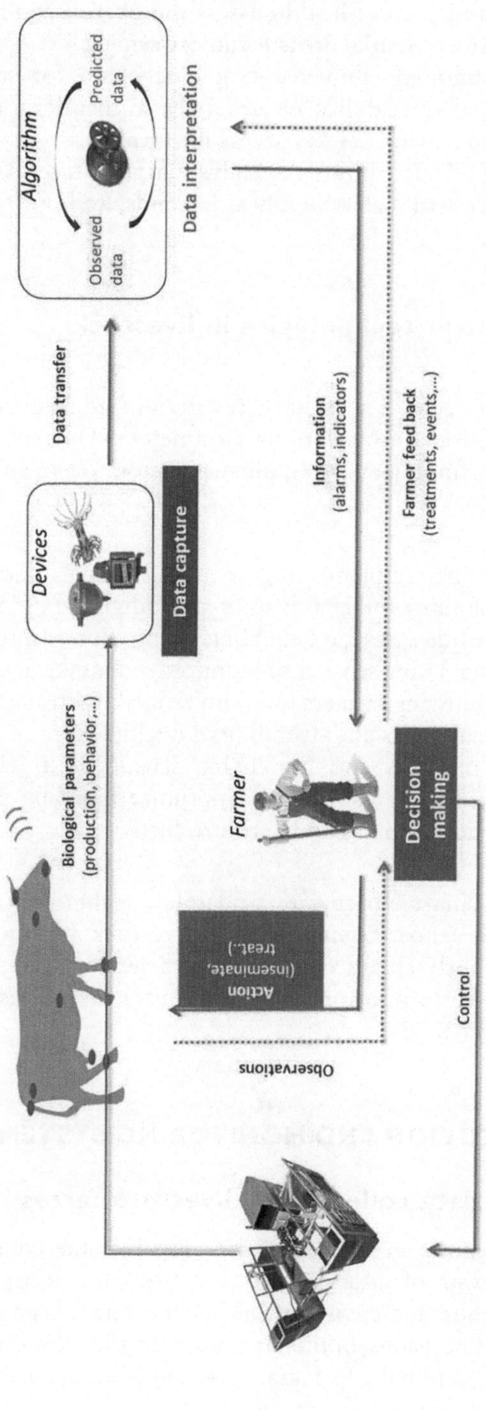


*Figure 1.3* Precision livestock farming.

**Source: Kleen and Guatteo, 2023.**

- **Drones and satellite imaging:** Utilizing aerial and satellite imagery to monitor crop and pasture health, assess the overall condition of the farm, and identify potential areas for improvement.
- **Blockchain technology:** Implementing blockchain for supply chain transparency and traceability, ensuring the authenticity of livestock products and improving market access for farmers.
- **AI:** Integrating AI algorithms to analyse large datasets and provide insights into livestock behaviour, health trends, and optimal management practices.

### 1.3.3 IoT and sensor technologies in livestock farming

The IoT and sensor technologies have revolutionized livestock farming by providing real-time data on various parameters. This subsection will explore how these technologies are employed in monitoring and managing livestock:

- **Health monitoring:** Implementing wearable devices and sensors to continuously monitor the health status of individual animals, detecting early signs of diseases, and enabling prompt intervention.
- **Precision feeding:** Using sensors to monitor feed intake and nutritional requirements, allowing farmers to optimize feeding strategies based on individual animal needs and overall herd health.
- **Reproduction management:** Employing sensors to track reproductive cycles, detect heat periods, and optimize breeding practices for improved genetic outcomes and herd productivity.

The integration of smart agriculture principles, technological advancements, and IoT and sensor technologies in livestock farming holds the potential to significantly enhance economic efficiency through improved productivity, resource utilization, and data-informed decision-making (Karunathilake *et al.*, 2023).

## 1.4 DATA COLLECTION AND MONITORING SYSTEMS

### 1.4.1 Real-time data collection in livestock farms

Real-time data collection in livestock farms involves the continuous and instantaneous gathering of information related to various aspects of livestock health, behaviour, and environmental conditions. This data is crucial for making informed decisions, optimizing resource allocation, and ensuring the well-being of the animals. In Figure 1.4, the components of real-time data collection in livestock farms are depicted.

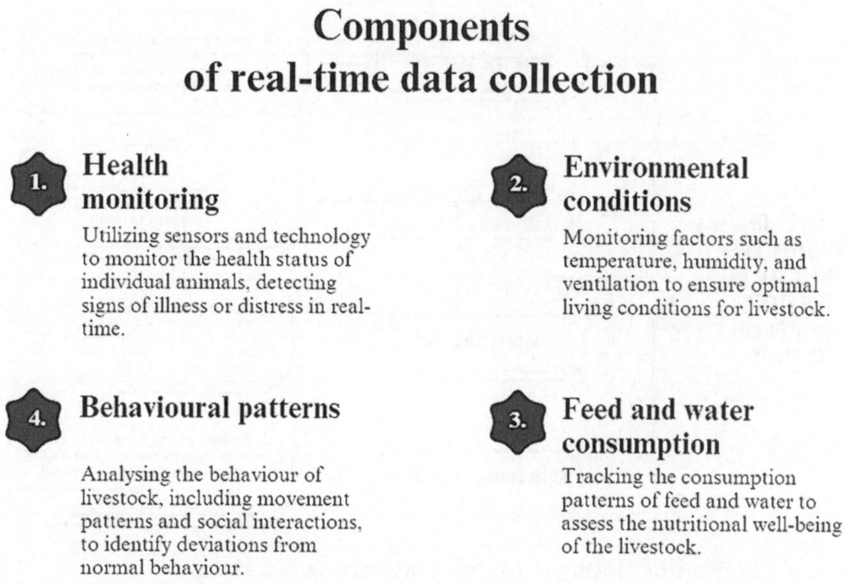


*Figure 1.4* Components of real-time data collection in livestock farms.

Real-time data collection provides farmers with immediate insights, allowing for timely interventions and proactive management strategies to enhance productivity and animal welfare (da Rosa Righi *et al.*, 2020).

### 1.4.2 Sensors and IoT devices for livestock monitoring

The integration of sensors and IoT devices has revolutionized livestock monitoring by providing a network of interconnected devices that collect, transmit, and analyse data. Figure 1.5 illustrates the utilization of various types of sensors in livestock monitoring systems.

The deployment of these sensors enables real-time data collection, creating a wealth of information that can be analysed to make data-driven decisions, predict potential issues, and optimize overall livestock management (Maduranga and Abeysekera, 2020).

### 1.4.3 Data storage and management solutions

Efficient data storage and management are critical components of a robust livestock monitoring system. The vast amount of data generated by sensors and monitoring devices requires secure and scalable solutions. Key considerations for data storage and management in livestock production include:

- **Cloud-based platforms**: Leveraging cloud computing for storing and processing data, providing accessibility, scalability, and real-time analytics from anywhere with an internet connection.

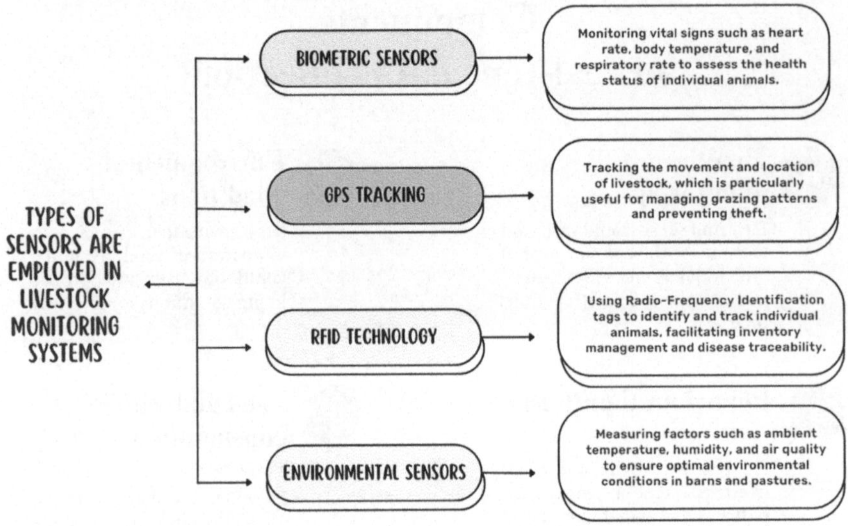


*Figure 1.5* Types of sensors in livestock monitoring systems.

- **Data security:** Implementing robust security measures to protect sensitive information, ensuring the confidentiality and integrity of livestock data.
- **Integration with farm management systems:** Seamless integration of livestock monitoring data with broader farm management systems to provide a holistic view of the entire agricultural operation.
- **Data analytics and visualization:** Employing data analytics tools to extract meaningful insights from the collected data, enabling farmers to make informed decisions. Visualization tools can also help in presenting complex data in a comprehensible manner.

The effective storage and management of data not only ensure the integrity of information but also lay the foundation for predictive analytics and decision support systems, ultimately contributing to economic efficiency in livestock production and smart agriculture (Mishra and Mishra, 2023).

The integration of advanced data collection and monitoring systems in livestock production is instrumental in enhancing economic efficiency through informed decision-making, proactive management, and optimization of resources in smart agriculture.

## 1.5 DATA ANALYSIS AND PREDICTIVE MODELLING

### 1.5.1 Importance of data analytics in livestock farming

a. Precision livestock farming:

- **Health monitoring:** Data analytics allows farmers to monitor the health of individual animals or the entire herd. Sensors and monitoring devices can collect data on vital signs, behaviour patterns, and other indicators, helping farmers detect signs of illness or distress early on.
- **Feed efficiency:** Analysing data on feed consumption, weight gain, and other parameters enables farmers to optimize feeding practices. This can lead to improved feed efficiency, reduced costs, and better overall animal performance.

b. **Resource management:**
- **Water and energy efficiency:** Analytics can help farmers monitor and manage water and energy usage on the farm. This is crucial for sustainability and cost-effectiveness.
- **Land use optimization:** By analysing data on grazing patterns and land utilization, farmers can optimize pasture management, ensuring that the available land is used efficiently and sustainably.

c. **Reproduction management:**
- **Breeding programmes:** Data analytics assists in the selection of breeding pairs based on genetic data, improving desirable traits in the herd. This can lead to enhanced productivity, disease resistance, and other beneficial characteristics.
- **Fertility monitoring:** Monitoring reproductive data allows farmers to identify the most fertile periods in animals, helping to optimize breeding programmes and increase overall reproductive success.

d. **Supply chain optimization:**
- **Inventory management:** Data analytics helps in managing and optimizing inventories of feed, medications, and other resources. This ensures that farmers have the right resources at the right time, minimizing waste and costs.
- **Market trends:** Analysing market trends and consumer preferences helps farmers make informed decisions about what products to produce, ensuring they meet market demand.

e. **Disease prevention and control:**
- **Early detection:** Data analytics aids in the early detection of disease outbreaks by monitoring changes in behaviour, vital signs, or other relevant parameters. This enables farmers to implement preventive measures promptly, reducing the impact of diseases on the herd.
- **Biosecurity:** Analytics can help farmers assess and improve biosecurity measures by identifying potential risk factors and implementing strategies to minimize the spread of diseases.

f. **Compliance and record keeping:**
- **Regulatory compliance:** Data analytics can assist farmers in maintaining accurate records to comply with regulations and standards in the livestock industry.

- **Traceability:** Having detailed data records allows for traceability in the supply chain, which is essential for quality control, food safety, and meeting consumer expectations.

### 1.5.2 Machine learning and AI for livestock prediction

Machine learning algorithms and AI applications are crucial components of livestock prediction systems. These technologies analyse large datasets, learn from patterns, and make predictions or recommendations, contributing to smart decision-making in livestock farming. In Figure 1.6, the applications and advantages of machine learning and AI for livestock prediction are illustrated.

### 1.5.3 Predictive models for livestock health and productivity

Predictive models for livestock health and productivity play a crucial role in modern agriculture, helping farmers make informed decisions to optimize animal well-being and production efficiency. Here are several types of predictive models used in this context:

a. **Disease prediction models:**
   - **Machine learning algorithms:** These models analyse historical health data, environmental conditions, and other relevant factors to predict the likelihood of disease outbreaks. They can help

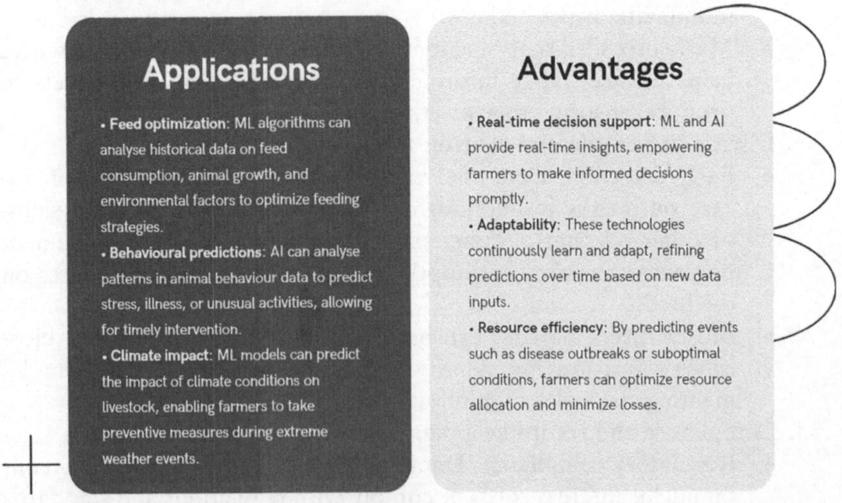


*Figure 1.6* Applications and advantages of machine learning and artificial intelligence for livestock prediction.

farmers take preventive measures, such as adjusting feeding practices or implementing biosecurity measures.

- **Epidemiological models:** These models simulate the spread of diseases within a population, considering factors like animal movement, contact patterns, and environmental conditions. They can aid in developing effective disease control strategies.

b. **Nutritional models:**
   - **Feed efficiency models:** These models predict the most efficient and cost-effective diet formulations for livestock based on factors like age, weight, and nutritional requirements. This helps farmers optimize feed usage and improve overall animal health.
   - **Metabolic models:** Predictive models can simulate metabolic processes in animals, helping farmers understand nutrient utilization and identify potential issues related to diet.

c. **Reproduction models:**
   - **Fertility prediction models:** These models use data on factors such as animal age, health status, and environmental conditions to predict reproductive success. This information can be crucial for optimizing breeding programmes and managing herd dynamics.
   - **Heat stress prediction models:** Livestock productivity can be affected by environmental factors like heat stress. Predictive models can help farmers anticipate periods of heat stress, allowing for proactive measures such as adjusting ventilation or providing cooling facilities.

d. **Behavioural models:**
   - **Animal welfare models:** These models predict and assess indicators of animal well-being based on behaviour, environmental conditions, and other factors. Monitoring behavioural patterns can help identify stress or discomfort in livestock.

e. **Environmental impact models:**
   - **Carbon footprint models:** Predictive models can estimate the environmental impact of livestock farming, helping farmers adopt sustainable practices and reduce greenhouse gas emissions.

f. **IoT and sensor-based models:**
   - **Wearable technology models:** The integration of sensors and wearable devices on livestock can provide real-time data on health parameters. Predictive models can analyse this data to identify early signs of illness or stress.

g. **Market and economic models:**
   - **Market price prediction models:** Predictive analytics can assist farmers in making strategic decisions related to the timing of livestock sales based on market trends and price predictions.

Data analysis, machine learning, and predictive modelling are integral components of smart agriculture, particularly in the context of livestock production. These technologies not only enhance economic efficiency but also contribute to sustainable and responsible farming practices (Yaqot and Menezes, 2021).

## 1.6 BENEFITS OF LIVESTOCK PRODUCTION MONITORING AND PREDICTION

### 1.6.1 Improved livestock health and well-being

Livestock production monitoring and prediction play a crucial role in enhancing the health and well-being of animals in smart agriculture. The integration of advanced technologies such as sensors, wearables, and IoT devices allows real-time tracking of various health parameters of livestock. This proactive approach enables farmers to identify and address health issues at an early stage, preventing the spread of diseases within the herd (Zaman, 2023).

- **Disease detection and prevention:** Continuous monitoring helps in the early detection of diseases and abnormal behaviour patterns. Predictive analytics, based on historical data and machine learning algorithms, can forecast potential health risks, allowing farmers to implement preventive measures and reduce the need for reactive interventions.
- **Precision nutrition:** Monitoring tools can assess the nutritional needs of individual animals, facilitating precision feeding programmes. This not only improves the overall health of the livestock but also optimizes feed utilization, reducing the risk of overfeeding or underfeeding.
- **Reproductive health management:** Predictive analytics can be employed to forecast optimal breeding times and identify potential fertility issues. This aids in better reproductive health management, leading to improved breeding success rates and healthier offspring.
- **Stress reduction:** Monitoring environmental factors such as temperature, humidity, and noise levels helps identify stressors that may affect livestock. By addressing these factors, farmers can create more comfortable and stress-free conditions for the animals, promoting better overall well-being.

### 1.6.2 Increased productivity and efficiency

The implementation of livestock production monitoring and prediction systems contributes significantly to enhancing productivity and efficiency in smart agriculture.

- **Optimized feeding programmes:** Continuous monitoring allows for the adjustment of feeding schedules and composition based on individual and herd requirements. This optimization leads to improved weight gain, faster growth rates, and overall better productivity.

- **Efficient resource allocation:** Predictive analytics can forecast trends in resource utilization, enabling farmers to allocate resources such as water, feed, and space more efficiently. This not only reduces waste but also ensures that the resources are used where they are most needed.
- **Early intervention for production issues:** Monitoring systems can detect anomalies in production patterns, signalling potential issues in advance. This allows farmers to take corrective actions promptly, minimizing disruptions and optimizing production efficiency.

### 1.6.3 Cost reduction and resource optimization

Livestock production monitoring and prediction contribute to cost reduction and the efficient use of resources, promoting economic sustainability in smart agriculture.

- **Optimized veterinary care:** By identifying health issues early on, farmers can optimize veterinary care, preventing the spread of diseases and reducing the need for extensive and costly treatments. This proactive approach leads to overall cost savings in healthcare for the livestock.
- **Reduced feed and resource waste:** Precision monitoring helps tailor feeding programmes to the specific needs of each animal, reducing the risk of overfeeding. Additionally, efficient resource allocation based on predictive analytics minimizes waste and ensures that resources are utilized optimally, reducing overall production costs.
- **Improved reproductive efficiency:** Predictive analytics in reproductive health management not only increases breeding success rates but also reduces the costs associated with unsuccessful breeding attempts. Efficient reproduction contributes to maintaining a healthy and productive herd, ultimately reducing the economic burden on farmers.

Benefits of livestock production monitoring and prediction in smart agriculture extend to improved health and well-being of livestock, increased productivity, and significant cost reductions through efficient resource utilization and proactive management practices. These advancements contribute to the overall economic efficiency and sustainability of livestock farming operations (Mitra *et al.*, 2022).

## 1.7 CHALLENGES AND LIMITATIONS

Efforts to enhance economic efficiency through livestock production monitoring and prediction in smart agriculture are met with several challenges and limitations. These hurdles range from technical obstacles in implementing monitoring systems to concerns related to data privacy, security, and economic and infrastructure barriers.

### 1.7.1 Technical challenges in implementing monitoring systems

The implementation of monitoring systems in smart agriculture faces several technical challenges that can impede their effectiveness. These challenges include:

- **Connectivity issues:** One of the primary technical challenges is ensuring consistent and reliable connectivity in remote agricultural areas. Limited internet access or unstable networks can hinder the real-time transmission of data from monitoring devices, affecting the accuracy and timeliness of livestock production information.
- **Sensor accuracy and reliability:** The precision and reliability of sensors used in monitoring systems are crucial for accurate data collection. Technical limitations in sensor technology, such as calibration issues, signal interference, or environmental factors, may lead to inaccurate readings, affecting the quality of predictions and decision-making.
- **Integration of diverse technologies:** Smart agriculture relies on the integration of various technologies, including IoT devices, machine learning algorithms, and cloud computing. Ensuring seamless interoperability among these technologies presents a challenge, as different components may have diverse technical requirements and compatibility issues.

### 1.7.2 Data privacy and security concerns

As the agriculture sector becomes more data-driven, concerns about data privacy and security become paramount. Farmers and stakeholders must address the following issues:

- **Confidentiality of farm data:** Farmers generate and share sensitive data related to livestock health, production metrics, and operational practices. Ensuring the confidentiality of this information is crucial to prevent unauthorized access and potential misuse by competitors or malicious entities.
- **Cybersecurity risks:** Smart agriculture systems are vulnerable to cyber threats, including data breaches, hacking attempts, and malware attacks. Securing the entire data ecosystem, from on-farm devices to cloud storage, is essential to protect against potential disruptions and unauthorized access.

### 1.7.3 Economic and infrastructure barriers

The adoption of monitoring and prediction systems in smart agriculture is hindered by economic and infrastructure barriers that limit accessibility and affordability.

Addressing these challenges and limitations is imperative for the successful integration of livestock production monitoring and prediction in smart agriculture, ultimately contributing to enhanced economic efficiency in the sector.

## 1.8 FUTURE TRENDS AND INNOVATIONS

In the rapidly evolving landscape of smart agriculture, the future trends and innovations in livestock production monitoring and prediction are essential to ensuring economic efficiency. This section explores some of the emerging technologies, the integration of blockchain and AI, and considerations for sustainability and the environment (Abobatta, 2021; Moysiadia *et al.*, 2021).

### 1.8.1 Emerging technologies in livestock farming

Livestock farming is witnessing a transformation through the adoption of cutting-edge technologies that enhance efficiency and productivity. Some emerging technologies include:

- **IoT:** IoT devices such as sensors and wearables play a crucial role in monitoring the health and behaviour of livestock. These devices collect real-time data on vital parameters like body temperature, heart rate, and movement patterns, enabling farmers to make data-driven decisions.
- **Satellite imaging and drones:** Satellite imagery and drones provide a bird's eye view of the entire farm, helping farmers monitor the condition of pastures, detect anomalies, and optimize the allocation of resources. This technology aids in efficient herd management and early detection of potential issues.
- **Robotics:** Autonomous robots are being deployed for various tasks, including feeding, cleaning, and monitoring livestock. These robots contribute to labour efficiency and can operate around the clock, ensuring constant supervision of the animals.

- **Genomic technologies:** Advances in genomics enable farmers to select livestock with desirable traits, such as disease resistance, higher milk production, or improved meat quality. This targeted breeding approach enhances overall herd quality and productivity.

### 1.8.2 Integration of blockchain and AI

The integration of blockchain and AI brings transparency, traceability, and efficiency to livestock farming practices. Here's how these technologies are influencing the industry:

- **Blockchain for supply chain transparency:** Blockchain technology ensures the traceability of livestock products throughout the supply chain. From farm to table, consumers can verify the origin, quality, and conditions under which the livestock was raised. This transparency fosters trust and helps combat fraud.
- **Smart contracts for automated transactions:** Smart contracts, powered by blockchain, enable automated and transparent transactions. Farmers can enter into agreements with suppliers, distributors, and other stakeholders with confidence, as the terms and conditions are executed automatically when predefined criteria are met.
- **AI for predictive analytics:** AI, particularly machine learning algorithms, processes vast amounts of data collected from various sources to predict trends and optimize farming practices. Predictive analytics can help farmers anticipate disease outbreaks, optimize feed formulations, and make informed decisions for efficient herd management.

### 1.8.3 Sustainability and environmental considerations

As agriculture faces increasing scrutiny regarding its environmental impact, the livestock sector is exploring sustainable practices. Key considerations include:

- **Precision farming:** Precision farming techniques involve the precise application of resources such as water, fertilizers, and feed. This approach minimizes waste, reduces environmental impact, and enhances resource use efficiency.
- **Carbon footprint reduction:** Livestock farming contributes to greenhouse gas emissions, and efforts are being made to reduce carbon footprints. This includes the adoption of alternative feed sources, improved waste management, and the implementation of sustainable land management practices.
- **Circular economy practices:** Embracing a circular economy involves recycling and reusing resources within the farming system. For example, converting animal waste into fertilizers or energy contributes to reducing environmental impact and promoting sustainability.

Future of livestock production monitoring and prediction in smart agriculture lies in the continuous integration of emerging technologies, the synergistic relationship between blockchain and AI, and a steadfast commitment to sustainability and environmental responsibility. These trends collectively contribute to achieving economic efficiency while addressing the challenges of the modern agricultural landscape (Gusev *et al.*, 2019).

## 1.9 RECOMMENDATIONS AND BEST PRACTICES

Efficient implementation of livestock monitoring and prediction systems in smart agriculture requires careful consideration of various factors (Mishra and Singh, 2023). This section outlines key recommendations and best practices to ensure economic efficiency and successful integration into the agricultural sector.

### 1.9.1 Guidelines for implementing livestock monitoring and prediction systems

To effectively implement livestock monitoring and prediction systems, farmers should follow these guidelines:

- **Define clear objectives and goals:** Before investing in the technology, farmers should clearly outline their objectives for implementing livestock monitoring and prediction systems. What specific improvements do they hope to achieve, such as increased milk production, reduced feed costs, or improved animal health? Having clear goals will help farmers make informed decisions about the type of technology to implement and how to measure the system's success.
- **Choose the right technology:** The choice of technology depends on several factors, including the type of livestock being monitored, the farm's infrastructure, and the farmer's technical expertise. Sensors, cameras, and software are among the various technologies available for livestock monitoring and prediction. Farmers should consider factors such as cost, ease of installation, and data compatibility when making their decision.
- **Collect high-quality data:** The quality of the data collected by the system is crucial for accurate predictions. Farmers should ensure that sensors are properly calibrated and maintained, and that data collection protocols are standardized to minimize errors. Additionally, collecting data from a representative sample of the livestock population is essential to ensure that the system's predictions are generalizable.
- **Develop robust prediction models:** Prediction models are at the heart of livestock monitoring and prediction systems. These models analyse the collected data to identify patterns and predict future outcomes, such as

animal health, milk production, or reproduction rates. Farmers should choose prediction models that are well suited for their data and their goals. Machine learning algorithms, such as linear regression, support vector machines, and neural networks, are often used for livestock prediction.

- **Integrate the system with existing workflows:** The livestock monitoring and prediction system should seamlessly integrate with the farmer's existing livestock management practices. This includes data management, decision support tools, and communication channels. Seamless integration ensures that the system is not a burden on the farmer's daily operations but rather a valuable asset that enhances efficiency.
- **Train staff on system usage:** Effective utilization of livestock monitoring and prediction systems requires proper training for farm staff. Farmers should provide comprehensive training on the system's operation, data interpretation, and decision-making processes. This ensures that staff can confidently use the system to optimize livestock management practices.
- **Monitor system performance and make adjustments:** Livestock monitoring and prediction systems are not static; they require ongoing monitoring and adjustments to maintain their accuracy and effectiveness. Farmers should regularly review the system's performance, evaluate its predictions against actual outcomes, and make adjustments to the system's parameters as needed.

### 1.9.2 Training and skill development for farmers

a. Tailored training programmes:
   - Develop training programmes customized to the needs and literacy levels of the farmers.
   - Include practical sessions for hands-on experience in using monitoring and prediction systems.

b. Digital literacy initiatives:
   - Collaborate with educational institutions and non-governmental organizations (NGOs) to promote digital literacy among farmers.
   - Provide resources and workshops to enhance farmers' understanding of smart agriculture technologies.

c. Localized support centres:
   - Establish support centres in rural areas to provide immediate assistance to farmers.
   - Ensure the availability of knowledgeable personnel to address technical issues and queries.

d. Incentive programmes:
   - Introduce incentive programmes to encourage farmers to adopt and excel in using monitoring systems.
   - Recognize and reward farmers who demonstrate exceptional proficiency in leveraging smart agriculture technologies.

### 1.9.3 Government and industry support in India

The Indian government has recognized the importance of the livestock sector and has implemented various initiatives to promote its growth and development. These initiatives focus on improving productivity, enhancing market access, and strengthening infrastructure. Some government programmes include:

- **National Livestock Mission (NLM):** The NLM aims to enhance livestock productivity, increase producer incomes, and ensure food security. It provides financial support for various activities, including breed improvement, animal health, feed and fodder development, and infrastructure development.
- **Livestock Health and Disease Control Programme (LH&DCP):** The LH&DCP focuses on controlling and eradicating animal diseases, such as foot and mouth disease, brucellosis, and tuberculosis. It provides funding for vaccination, diagnostic testing, and extension services.
- **Dairy Processing and Infrastructure Development Fund (DPIDF):** The DPIDF provides financial assistance for the establishment and modernization of dairy processing facilities. It also supports the development of milk collection centres and refrigerated transport infrastructure.
- **Livestock Export Development Authority (LEDA):** LEDA promotes the export of livestock products by providing financial assistance, market intelligence, and training to exporters. It also organizes trade fairs and exhibitions to showcase Indian livestock products to international buyers.
- **National Livestock Policy 2013:** The National Livestock Policy 2013 outlines a comprehensive strategy for the development of the livestock sector. It aims to increase productivity, enhance market access, and ensure sustainability.

In addition to government initiatives, the Indian industry is playing a significant role in promoting the adoption of smart agriculture technologies in livestock production. Several private companies are developing and deploying solutions for livestock monitoring, prediction, and management. These solutions are helping farmers to improve productivity, reduce costs, and enhance animal health (Rehman *et al.*, 2022; Sinha and Dhanalakshmi, 2022).

Here are some specific examples of government and industry support for smart agriculture technologies in livestock production in India:

- The Indian Council of Agricultural Research (ICAR) has developed a mobile app called "e-Gopala" that provides farmers with information on livestock management practices, disease prevention, and market prices.

- Tata Consultancy Services (TCS) has partnered with the National Dairy Development Board (NDDB) to develop a smart dairy management system that uses IoT sensors and data analytics to monitor animal health, milk production, and feed consumption.
- Microsoft India has launched the "Microsoft Azure FarmBeats" platform, which provides a cloud-based solution for livestock monitoring and management.
- SAP India has partnered with the World Wildlife Fund (WWF) to develop a mobile app called "Smart Livestock Management" that helps farmers to reduce their environmental impact.

These are just a few examples of the many initiatives that are being undertaken to promote smart agriculture technologies in livestock production in India. These technologies have the potential to significantly improve the efficiency and sustainability of the sector, while also enhancing the incomes of livestock farmers.

The Indian government and industry are committed to supporting the development and adoption of smart agriculture technologies in livestock production. These technologies have the potential to transform the sector and make it a more efficient, sustainable, and profitable industry (Tao *et al.*, 2021; Wang *et al.*, 2021).

## 1.10 CONCLUSION

Livestock plays a pivotal role in agriculture, serving as a major component that contributes significantly to the economic landscape. Livestock production holds economic significance, providing essential resources such as meat, dairy, and wool. However, this sector faces multifaceted challenges, encompassing issues related to health, nutrition, and environmental impact. The integration of smart agriculture and technology emerges as a transformative solution to address these challenges. Smart agriculture, with its emphasis on harnessing technological advancements, introduces innovative concepts that redefine traditional farming practices. Technology, particularly the IoT and sensor technologies, plays a crucial role in enhancing livestock farming efficiency by providing real-time data on various parameters. Efficient data collection and monitoring systems become integral components of smart livestock farming. The deployment of sensors and IoT devices facilitates continuous monitoring, ensuring that farmers have access to real-time information. Robust data storage and management solutions further streamline the handling of large datasets, paving the way for informed decision-making. The significance of data analytics in livestock farming cannot be overstated. Leveraging machine learning and AI for predictive modelling enables farmers to anticipate livestock health issues and optimize productivity. The benefits are manifold, ranging from improved livestock health and well-being to increased overall efficiency, cost reduction, and resource optimization. However, the implementation of monitoring systems is not without challenges. Technical hurdles, data privacy concerns, and economic barriers pose obstacles to widespread adoption. Despite these limitations, the future holds promise with emerging technologies such as blockchain and increased emphasis on sustainability and environmental considerations. To navigate this evolving landscape, recommendations and best practices are crucial. Guidelines for implementing monitoring and prediction systems, along with targeted training and skill development for farmers, are essential. Government and industry support are imperative to foster the integration of these technologies, ensuring a sustainable and efficient future for livestock production.

## REFERENCES

Abdalla, Z. F. and El-Ramady, H., "Applications and challenges of smart farming for developing sustainable agriculture," *Environ. Biodivers. Soil Sec.*, vol. 6, no. 2022, pp. 81–90, 2022 [doi:10.21608/jenvbs.2022.135889.1175].

Abdullahi, H. S. et al., "Technology impact on agricultural productivity: A review of precision agriculture using unmanned aerial vehicles" in *Wireless and Satellite Systems*. Revised Selected Papers 7: *7th International Conference, WiSATS 2015*, Bradford, UK, July 6–7, 2015. Springer International Publishing, pp. 388–400.

Abobatta, W. F., "Precision agriculture: A new tool for development" in *Precision Agriculture Technologies for Food Security and Sustainability*. IGI Global, 2021, pp. 23–45 [doi:10.4018/978-1-7998-5000-7.ch002].

Alonso, R. S. et al., "An intelligent Edge-IoT platform for monitoring livestock and crops in a dairy farming scenario," *Ad Hoc Netw.*, vol. 98, p. 102047, 2020 [doi:10.1016/j.adhoc.2019.102047].

Banhazi, T. M. et al., "Precision livestock farming: An international review of scientific and commercial aspects," *Int. J. Agric. Biol. Eng.*, vol. 5, no. 3, pp. 1–9, 2012.

Bhattacharyya, P. et al., "Livestock and aquaculture management for climate-smart agriculture" *Climate Smart Agriculture: Concepts, Challenges, and Opportunities*, 2020, pp. 113–127 [doi:10.1007/978-981-15-9132-7_8].

Chatterjee, P. S. et al., "LiveCare: An IoT-based healthcare framework for livestock in smart agriculture," *IEEE Trans. Con. Electron.*, vol. 67, no. 4, pp. 257–265, 2021 [doi:10.1109/TCE.2021.3128236].

Choi, S. W. and Shin, Y. J., "Role of smart farm as a tool for sustainable economic growth of Korean agriculture: Using input–output analysis," *Sustainability*, vol. 15, no. 4, p. 3450, 2023 [doi:10.3390/su15043450].

da Rosa Righi, R. et al., "Towards combining data prediction and internet of things to manage milk production on dairy cows," *Comput. Electron. Agric.*, vol. 169, p. 105156, 2020 [doi:10.1016/j.compag.2019.105156].

Dahane, A. et al., "An IoT low-cost smart farming for enhancing irrigation efficiency of smallholders farmers," *Wirel. Personal Commun.*, vol. 127, no. 4, pp. 3173–3210, 2022 [doi:10.1007/s11277-022-09915-4].

De Alwis, S. D. et al., "A survey on smart farming data, applications and techniques," *Comput. Ind.*, vol. 138, p. 103624, 2022 [doi:10.1016/j.compind.2022.103624].

Gupta, A. and Nahar, P., "Classification and yield prediction in smart agriculture system using IoT," *J. Ambient Intell. Hum. Comput.*, vol. 14, no. 8, pp. 10235–10244, 2023 [doi:10.1007/s12652-021-03685-w].

Gupta, M. et al., "Security and privacy in smart farming: Challenges and opportunities," *IEEE Access*, vol. 8, pp. 34564–34584, 2020 [doi:10.1109/ACCESS.2020.2975142].

Gusev, A. S. et al., "An analysis of research areas in precision agriculture," *Int. Trans. J. Eng. Manag. Appl. Sci. Technol.*, vol. 10, no. 10, pp. 1–10, 2019.

Hedley, C., "The role of precision agriculture for improved nutrient management on farms," *J. Sci. Food Agric.*, vol. 95, no. 1, pp. 12–19, 2015 [doi:10.1002/jsfa.6734].

Higgins, S. et al., "Improving productivity and increasing the efficiency of soil nutrient management on grassland farms in the UK and Ireland using precision agriculture technology," *Eur. J. Agron.*, vol. 106, pp. 67–74, 2019 [doi:10.1016/j.eja.2019.04.001].

Jaiswal, S. P. et al., "Internet of Things (IoT) for smart agriculture and farming in developing nations," *Int. J. Sci. Technol. Res.*, vol. 8, no. 12, pp. 1049–1056, 2019.

Javaid, M. et al., "Enhancing smart farming through the applications of Agriculture 4.0 technologies," *Int. J. Intell. Netw.*, vol. 3, pp. 150–164, 2022 [doi:10.1016/j.ijin.2022.09.004].

Kakamoukas, G. et al., "Towards climate smart farming—A reference architecture for integrated farming systems" in *Telecom* (Vol. 2, No. 1, pp. 52-74). MDPI, vol. 2, no. 1, 52–74, 2021, Febr. [doi:10.3390/telecom2010005].

Kumar, N. et al., "A study on costs and returns of paddy cultivation in Ambedkar Nagar district of Uttar Pradesh". *Int. J. Stat. Appl. Math.*, SP-8, vol. 3, pp. 107–111, 2023.

Kleen, J. L. and Guatteo, R., "Precision livestock farming: What does it contain and what are the perspectives?," *Animals*, vol. 13, no. 5, pp. 779, 2023.

Lovarelli, D. et al., "A review on dairy cattle farming: Is precision livestock farming the compromise for an environmental, economic and social sustainable production?," *J. Cleaner Prod.*, vol. 262, p. 121409, 2020 [doi:10.1016/j.jclepro.2020.121409].

Maduranga, M. W. P. and Abeysekera, R., "Machine learning applications in IoT based agriculture and smart farming: A review," *Int. J. Eng. Appl. Sci. Technol.*, vol. 4, no. 12, pp. 24–27, 2020.

Madushanki, A. A. R. et al., "Adoption of the Internet of Things (IoT) in agriculture and smart farming towards urban greening: A review," *IJACSA*, vol. 10, no. 4, pp. 11–28, 2019 [doi:10.14569/IJACSA.2019.0100402].

Mekonnen, Y. et al., "Review—Machine learning techniques in wireless sensor network based precision agriculture," *J. Electrochem. Soc.*, vol. 167, no. 3, p. 037522, 2020 [doi:10.1149/2.0222003JES].

Mishra, H. and Mishra, D., Eds., "Artificial intelligence and machine learning in agriculture: Transforming farming systems" in *Research Trends in Agriculture Science*, vol. I. Bhumi Publishing, 2023, pp. 1–16.

Mishra, H. and Singh, M., "Revolutionizing agriculture: Empowering farmers through technological adoption for increased productivity and profitability," *The Agric. Mag.*, vol. 2, no. 8, pp. 17–22, 2023.

Mishra, S. and Sharma, S. K., "Advanced contribution of IoT in agricultural production for the development of smart livestock environments," *Internet Things*, vol. 22, p. 100724, 2023 [doi:10.1016/j.iot.2023.100724].

Mitra, A. et al., 2022, "Everything you wanted to know about smart agriculture," *arXiv Preprint ArXiv:2201.04754*.

Monteiro, A. et al., "Precision agriculture for crop and livestock farming—Brief review," *Animals (Basel)*, vol. 11, no. 8, p. 2345, 2021 [doi:10.3390/ani11082345].

Morrone, S. et al., "Industry 4.0 and precision livestock farming (PLF): An up to date overview across animal productions," *Sensors (Basel)*, vol. 22, no. 12, p. 4319, 2022 [doi:10.3390/s22124319].

Moysiadis, V. et al., "Smart farming in Europe," *Comput. Sci. Rev.*, vol. 39, p. 100345, 2021 [doi:10.1016/j.cosrev.2020.100345].

Neethirajan, S. and Kemp, B., "Digital livestock farming," *Sens. Bio Sens. Res.*, vol. 32, p. 100408, 2021 [doi:10.1016/j.sbsr.2021.100408].

Nishad, D. C. et al., Eds., 2023, "Towards sustainable agriculture: Mitigating the adverse effects of stubble burning in India," *Research Trends in Environmental Science*, vol. I. Bhumi Publishing.

O'Grady, M. J. and O'Hare, G. M. P., "Modelling the smart farm," *Inf. Process. Agric.*, vol. 4, no. 3, pp. 179–187, 2017 [doi:10.1016/j.inpa.2017.05.001].

Pathan, M. et al., "Artificial cognition for applications in smart agriculture: A comprehensive review," *Artif. Intell. Agric.*, vol. 4, pp. 81–95, 2020 [doi:10.1016/j.aiia.2020.06.001].

Perakis, K. et al., "CYBELE–Fostering precision agriculture & livestock farming through secure access to large-scale HPC enabled virtual industrial experimentation environments fostering scalable big data analytics," *Comput. Netw.*, vol. 168, p. 107035, 2020 [doi:10.1016/j.comnet.2019.107035].

Ramya, M. G. P. G. et al., "Environment change prediction to adapt climate-smart agriculture using big data analytics," *Int. J. Adv. Res. Comput. Eng. Technol. (IJARCET)*, vol. 4, no. 5, 2015.

Rehman, A. et al., "A revisit of internet of things technologies for monitoring and control strategies in smart agriculture," *Agronomy*, vol. 12, no. 1, p. 127, 2022 [doi:10.3390/agronomy12010127].

Said Mohamed, E. et al., "Smart farming for improving agricultural management," *Egypt. J. Remote Sens. Space Sci.*, vol. 24, no. 3, pp. 971–981, 2021 [doi:10.1016/j.ejrs.2021.08.007].

Schellberg, J. et al., "Precision agriculture on grassland: Applications, perspectives and constraints," *Eur. J. Agron.*, vol. 29, no. 2-3, pp. 59–71, 2008 [doi:10.1016/j.eja.2008.05.005].

Seelan, S. K. et al., "Remote sensing applications for precision agriculture: A learning community approach," *Remote Sens. Environ.*, vol. 88, no. 1-2, pp. 157–169, 2003 [doi:10.1016/j.rse.2003.04.007].

Sharma, A. et al., "Enabling smart agriculture by implementing artificial intelligence and embedded sensing," *Comput. Ind. Eng.*, vol. 165, p. 107936, 2022 [doi:10.1016/j.cie.2022.107936].

Singh, D. K. et al., "LoRa based intelligent soil and weather condition monitoring with internet of things for precision agriculture in smart cities," *IET Commun.*, vol. 16, no. 5, pp. 604–618, 2022 [doi:10.1049/cmu2.12352].

Sinha, B. B. and Dhanalakshmi, R., "Recent advancements and challenges of Internet of Things in smart agriculture: A survey," *Future Gener. Comput. Syst.*, vol. 126, pp. 169–184, 2022 [doi:10.1016/j.future.2021.08.006].
Su, Y. and Wang, X., "Innovation of agricultural economic management in the process of constructing smart agriculture by big data," *Sustain. Comput. Inform. Syst.*, vol. 31, p. 100579, 2021 [doi:10.1016/j.suscom.2021.100579].
Tao, W. et al., "Review of the internet of things communication technologies in smart agriculture and challenges," *Comput. Electron. Agric.*, vol. 189, p. 106352, 2021 [doi:10.1016/j.compag.2021.106352].
Tedeschi, L. O. et al., "Advancements in sensor technology and decision support intelligent tools to assist smart livestock farming," *J. Anim. Sci.*, vol. 99, no. 2, p. skab038, 2021 [doi:10.1093/jas/skab038].
Tiwari, A. K. et al., Eds., "Sustainable water management in agriculture: Irrigation techniques and water conservation" in *Research Trends in Agriculture Science*, vol. II. Bhumi Publishing, 2023, pp. 53–68.
Ukhurebor, K. E. et al., "Precision agriculture: Weather forecasting for future farming" in *Ai, Edge and IoT-Based Smart Agriculture*. Academic Press, 2022, pp. 101–121.
Vrchota, J. et al., "Precision agriculture technologies for crop and livestock production in the Czech Republic," *Agriculture*, vol. 12, no. 8, p. 1080, 2022 [doi:10.3390/agriculture12081080].
Wang, T. et al., "From smart farming towards unmanned farms: A new mode of agricultural production," *Agriculture*, vol. 11, no. 2, p. 145, 2021 [doi:10.3390/agriculture11020145].
Wolfert, S. et al., "Big data in smart farming–a review," *Agric. Syst.*, vol. 153, pp. 69–80, 2017 [doi:10.1016/j.agsy.2017.01.023].
Yaqot, M. and Menezes, B. C., "Unmanned aerial vehicle (UAV) in precision agriculture: Business information technology towards farming as a service" in *1st international conference on emerging smart technologies and applications (eSmarTA)*, vol. 2021. IEEE, 2021, Aug., pp. 1–7 [doi:10.1109/eSmarTA52612.2021.9515736].
Zaman, Q. U., "Precision agriculture technology: A pathway toward sustainable agriculture" in *Precision Agriculture*. Academic Press, 2023, pp. 1–17.
Zhang, F. and Cao, N., "Application and research progress of Geographic Information System (GIS) in agriculture" in *8th International Conference on Agro-Geoinformatics (Agro-Geoinformatics)*, vol. 2019. IEEE, 2019, Jul., pp. 1–5 [doi:10.1109/Agro-Geoinformatics.2019.8820476].

Chapter 2

# Smart crop management for increased yield

*Jyotirmay Mahapatra, Manas Mohanty, Debasis Gountia and Mrutyunjay Senapati*

## 2.1 INTRODUCTION

The need for both qualitative and quantitative agricultural products is rising. Between 2005 and 2050, the demand is expected to increase from 60% (Alexandratos & Bruinsma, 2012) to over 100% (Tilman et al., 2011). The quality of agricultural products is becoming increasingly essential as people lead healthier lifestyles and become more aware of their health. The growth of agriculture has been significantly impacted by climate change. Abiotic stresses on plants, such as drought, flooding, and temperature, are growing. In such unfavourable circumstances, the additional demand must be met with the limited resources that are available. There are few primary resources for agriculture, such as water and land. The environmental cost of resources such as electricity, seed, fertiliser, pesticides, weedicides, and fungicides is significant. The overall cost of the investment rises as input costs rise. The agricultural environment and farm produce are adversely affected by the high rate of pesticide applications. Considering the aforementioned factors, the farming sector must adopt new technologies.

Crop management includes different operations in agriculture, starting from land preparation to harvesting and residue management. Except these, decision-making is crucial in regard to each activity. Soil preparation involves activities such as tilling, levelling, and adding soil amendments to create a suitable environment for plant growth. Proper soil preparation ensures good seedbed conditions and nutrient availability. The timing and method of planting are critical factors in crop management. Farmers must consider factors such as climate, soil conditions, and the specific requirements of the crop to optimise germination and establishment. Efficient water management is essential for crop success. Crops require essential nutrients for growth. Fertilisers need to be applied based on soil nutrient analysis and the specific needs of the crops. This helps maintain soil fertility and ensures healthy plant development. Monitoring for pests and diseases is crucial to preventing or minimising crop damage. Weeds compete with crops for resources, so effective weed management is essential. This may involve mechanical cultivation, herbicide application, or other methods to suppress

DOI: 10.1201/9781003628392-2

weed growth. Regular monitoring of crop health, growth, and environmental conditions is vital. This can involve visual inspections, the use of sensors, satellite imagery, or other technologies to assess the status of the crops. Timing the harvest is critical to ensuring optimal yield and quality. Proper harvesting methods, such as using the right equipment and techniques, help prevent losses and maintain crop quality. Once crops are harvested, they need proper handling and storage to maintain quality. This includes activities like drying and storing crops to prevent spoilage and maintain market value. The handling of remnants should be determined appropriately, considering factors such as the type of crops, geographical location, and existing amenities.

Effectiveness within all these operations can be improved by reducing losses of resources and produce. Losses can be reduced by adopting site-specific management practices at an appropriate time. Site-specific management requires analysing land, crop, and environment and then doing the activities as per the spatial requirements based on the analysis. These analyses involve a huge amount of data from several affecting parameters. Manual observation, conventional techniques, and human decisions are not efficient. Continuous data collection, inclusion of all affecting parameters, development of models, application of advanced computing, communication, and actuation techniques in agriculture enable appropriate and efficient management in a mechanised way. Figure 2.1 presents data regarding the adaptation of site-specific management practices in developed countries. The necessity of cutting-edge technologies for data collection, including sensor nodes, agricultural robots, unmanned aerial vehicles (UAVs), and high-resolution satellite imagery, was stressed by the European Union (EU). These technologies are critical to European agriculture's future smart farming strategies, as outlined in the April 2019 agreement signed by 24 of the EU countries (Bacco et al., 2019).

To address significant challenges with food security, adaptation, and mitigation, climate-smart agriculture, or CSA, was created (Palombi and Sessa, 2013). Owing to its capacity to reduce greenhouse gas emissions, increase farm system resilience, and enhance food security, CSA has attracted a lot of attention, particularly in developing countries (Palombi and Sessa, 2013). A revolutionary development from precision farming, smart agriculture offers cutting-edge techniques for multifunctional farm management that are remotely supported.

## 2.2 SENSORS FOR DATA COLLECTION

Currently, developed countries like the United States, Japan, and Germany are leading other nations in the area of industrial processes and technologies. Soil, weather, water, and plant sensors are among the many diverse uses for agricultural sensors that are becoming more and more common. The

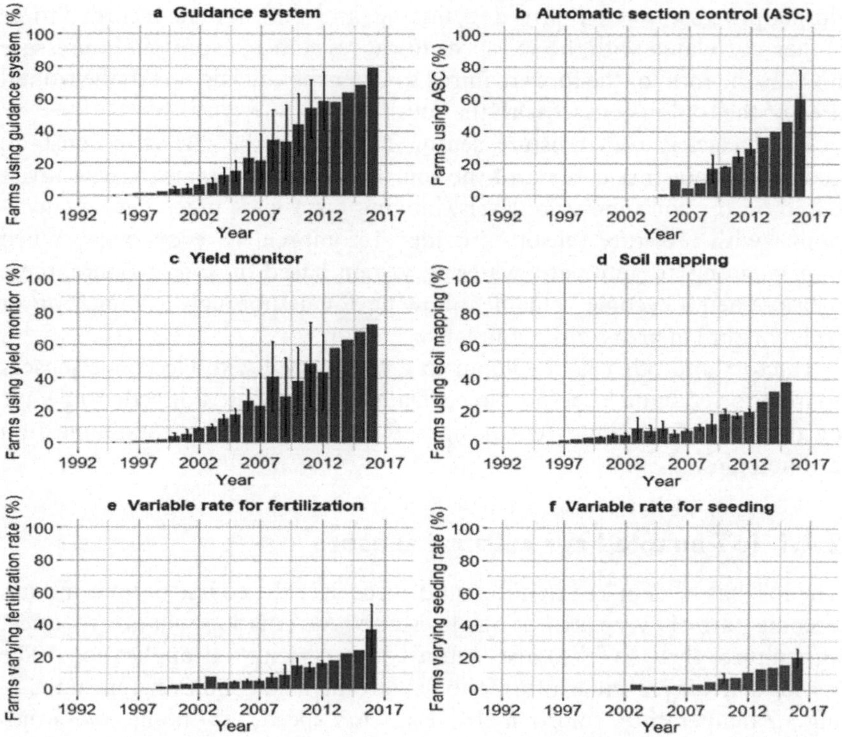


*Figure 2.1* Increasing adaptation rate of precision farming operations in developed countries.

**Nowak, 2021.**

sensors that identify different items offer significant assistance in gathering data on agricultural productivity.

A variety of optical sensors are employed to track multiple occurrences of light and analyse soil materials based on light properties. These sensors can be affixed to vehicles, spacecraft, drones, and robotic devices, collecting and scrutinising reflected data. Parameters such as clay content, organic matter, and humidity levels in the soil are all obtainable through these optical sensors.

Electrochemical sensors play a crucial role in collecting, processing, and mapping the chemical data of the soil. Typically placed on carefully crafted sleds, these sensors are used to determine soil pH and nutrient content. After this assessment, soil samples are transported to a laboratory for conventional testing procedures. An ion-selective electrode is a reliable method for accurately measuring pH and detecting specific ions like hydrogen, nitrate, and potassium.

For assessing mechanical resistance or soil compression, load cells are employed. These sensors move through the soil, measuring the force exerted during processes such as crushing, tearing, and shifting. The recorded force is then correlated with the soil mechanical resistance, expressed in pressure units as the ratio of the force required to enter the soil channel to the frontal area of the tool in contact with the soil.

The dielectric soil moisture sensor, relying on the dielectric constant, gauges the moisture content of the soil. This electrical characteristic varies based on the soil's moisture levels. Linking farm-wide precipitation checkpoints with moisture sensors provides a comprehensive overview. When implementing an automatic watering system based on soil moisture, these measurements become valuable, allowing for a thorough examination of soil moisture in areas with minimal vegetation.

Tractors and other field equipment are equipped with electronic sensors to monitor equipment performance. Data are distributed directly to people or transferred to computers using satellite and cellular communication technologies.

### 2.2.1 IoT-enabled agricultural sensors

The IoT has been linked to different devices over the course of time and has become part of every area of our lives now. Automation will greatly affect agriculture; thus, IoT has also entered into agricultural applications. The sensor delivers real-time information regarding field conditions, encompassing air temperature, chlorophyll levels, wind speed, dew point, and atmospheric pressure, among other data points.

### 2.2.2 Precision agriculture sensors

The total world population has been growing at an alarming rate over the past few decades, which has pushed farming technologies to become more sophisticated, competitive, large-scale, and optimised. Technology has made farming operations more productive than they were previously, which has increased the amount and quality of produce that farmers are able to obtain.

#### *2.2.2.1 Global Positioning System sensors*

Vehicle navigation systems have adopted GPS sensors, providing more precision. Auto-guided farming apps can minimise redundant operations and ultimately reduce the needed time to finish a task. Modern GPS makes it easier to track livestock because it can monitor the animals with just a button push, eliminating the need for manual tracking.

#### *2.2.2.2 Temperature sensor*

Temperature sensors aid in crop assessment. Temperature sensors oversee both the machinery responsible for harvesting the plants and the plants themselves during the harvesting process. These sensors guard against equipment overheating and catastrophic malfunctions.

#### 2.2.2.3 Accelerometer sensor

Accelerometers have a frequent application in drives and other dynamic components. Their key role is to detect small movements and rotations. Their primary responsibility is to predict when a potentially compromised component will require routine maintenance or replacement. Accelerometers show their versatility in a range of automated systems and tracking techniques. For example, the integration of a low-power accelerometer enables the effective monitoring of the state of an adjustable spray nozzle at the extremity of a fertilisation beam. When using autonomous drones for intelligent agricultural practices, advanced technology heavily depends on accelerometers and inertial measuring units (IMUs) to track motion, speed, collisions, and spatial positioning.

#### 2.2.2.4 Smart cameras used in agriculture

The use of smart cameras in smart agriculture applications has grown over time. Because of this, fertiliser and herbicides are dispensed automatically and precisely. This reduces chemical usage while increasing overall productivity through the use of chemicals.

In agriculture, controlling pests is one of the biggest challenges. Smart camera technology, enables farmers to identify pests in real time and keep a close eye on pest control efforts without endangering beneficial agricultural insects. Additionally, semi-legacy sensing devices like ambient light monitoring can be replaced by smart cameras, which allows for a more streamlined system with fewer components. Precision farming is an approach that involves applying certain inputs at different rates in order to maximise economic efficiency and minimise waste. Precision agriculture sensors make it easier to switch from traditional farming practices to precision agriculture. Among its many impressive features is the ability to generate and analyse precise data using a cloud-based platform. GeoPard Agriculture also provides intelligent surveying capabilities. With their ability to facilitate planning, execution, and adaptation based on available data, these tools are dependable aids for agricultural endeavours. Using GeoPard can provide the best answer and was essential to the founding of many agricultural-focused applications businesses.

A red, green, and blue, or RGB camera, can be used to take pictures for a variety of applications, including improving security, keeping an eye on plant health, and determining the best times to harvest crops. This entails processing the data using algorithms based on the images and comparing them to a reference crop harvesting image (Gagliardi et al., 2021). Hyperspectral cameras, which function in the visible, infrared, and ultraviolet light spectrums are also very important. Specialised farmers can now accurately monitor field conditions, crop productivity, and irrigation thanks to this technology. Each dataset from a hyperspectral camera goes through a filtration process to improve the information obtained by focusing light on particular colours or wavelengths (Ramirez et al., 2022).

Utilising thermal imaging proves to be an effective approach in determining crucial factors for irrigation scheduling, as substantiated by numerous studies. Key elements such as water stress, gas exchange rate, evapotranspiration rate, stomatal conductance, and stomata closure play a vital role in irrigation considerations. When a plant undergoes water stress, its stomata close, interrupting transpiration, leading to an increase in the plant's temperature and canopy temperature. Consequently, by examining stomatal responses, thermal remote sensing allows for the measurement of plant temperature, stomatal conductance, and evapotranspiration rate. A noteworthy advantage of thermal imaging over traditional thermometry lies in its ability to provide an average temperature value, whereas thermometry merely offers values for individual pixels within the sensor's field of vision. This distinction proves valuable in discerning between various elements, such as sunlight-exposed versus covered plant sections and distinguishing wet from dry soil surfaces. Recent advancements in thermal imaging, coupled with enhanced image processing and data analytics, contribute to mitigating crop water stress and facilitating improved irrigation scheduling.

#### 2.2.2.5 *Nitrogen, phosphorus, and potassium sensor*

The topsoil nitrogen, phosphorus, and potassium (NPK) sensing instrument aids in determining the soil's concentrations of potassium (K), phosphorus (P), and nitrogen (N). These three determine soil fertility and are necessary for the best possible plant growth. It assists in informing the farmer about the best times to apply fertiliser to the farm in order to preserve soil nutrients (Ratnaparkhi et al., 2020).

### 2.2.3 Advantages of sensor applications and limitations

Building well-predictive models with the aforementioned methods requires a vast amount of data. It is impossible to manually gather this much data. Accordingly, a variety of sensors, including those for temperature, pressure, humidity, moisture content, colour, sound, light, orientation, location, acceleration, distance, proximity, and chemical sensing, are used to collect data.

It is important to take into account the limits of agricultural sensors. First of all, there is a chance of measurement mistakes or inconsistencies due to variations in sensor accuracy and reliability. Second, in order to guarantee optimal operation, some sensors could need regular calibration or maintenance. Third, the capacity of sensors to offer thorough insights may be limited if specific environmental conditions or fluctuations are not precisely captured by them.

## 2.3 COMMUNICATION AND STORAGE

Large volumes of data are being employed in agricultural management, coinciding with the expansion of sensing systems for data collection, processing, and analysis. As a result, this reduces the 4G network's capacity to link distant components. Data processing and transfer have recently become simpler processes with the launch of the ultrafast 5G switch (Tang et al., 2021).

In recent years, communication and information technology have significantly advanced in tandem with the progress of smart systems. Over the past 10 years, advances in 3G, 4G, and NB-IoT wireless network technology have made it possible to link smart devices via the IoT and exchange data for precise assessment in the farm at a speed that is sufficient for information transmission and communication. But due to the huge increase in volume and quality of information, the ability of the 4G network to transmit data is less efficient than in the past. According to Sutton et al. (2018), the 5G network, which offers incredibly quick data transformation, is the advancement of fifth-generation communication. Since the upload and download speeds of the 4G network rise by over 100 times, the data transfer over the 5G network is quicker than other networks (Anand et al., 2020).

For the purpose of smart farming, a variety of communication protocols are commonly employed in farming. By applying these protocols, easier and more informed decisions can be made by farmers, which will enhance tracking crop growth. IEEE 802.11 Wi-Fi, LoraWan, 2G/3G/4 G-Mobile Communications Standards, Low Rate Wireless Personal Area Networks, WiMax, Bluetooth, ZigBee, and radio Frequency Identification (RFID) are the most popular wireless technologies. Table 2.1 provides a comparison between the aforementioned wireless protocols.

a. IEEE 802.11 wireless local area network (WLAN) technologies that consist of 802.11a, 802.11b, 802.11ac, 802.11g, and 802.11n are compiled into one specification. These standards all function at various frequencies, like 2.2 GHz, 2.4 GHz, 2.4/5 GHz, 5 GHz, and 60 GHz. These standards cover a bandwidth of 1 Mb/s to 7 Gb/s for data transfer. It can communicate at a distance of 20 to 100 metres (Lai et al., 2012). For smart agriculture, a Wi-Fi-based wireless sensor network has been considered in the IoT (Ahmad et al., 2018).

*Table 2.1* Comparison of communication technologies used in agricultural Internet of Things (IoT) (Xu et al., 2022)

| *Category* | *Wireless WAN (GPRS/4G/5G)* | *Wi-Fi* | *Bluetooth* | *ZigBee* | *Lora* |
|---|---|---|---|---|---|
| Use | Call, information | Media | Media, cable alterative | Monitoring, sensors | Data transparent transmission |
| Power usage | Elevated | Elevated | Diminutive | Diminutive | Diminutive |
| Communication range | >15 km | < 100 m | < 10 m | 10–100 m | 2–5 km in towns and 15 km in the outskirts |
| Benefit | Extensive coverage, excellent service quality | Great adaptability | Affordability | Efficient energy utilisation | Robust immunity with steady operation |

b. The long-range communication protocol LoraWan is being developed by the Lora TM Alliance. The major goal of this protocol is to secure interoperability across several operators (Dias et al., 2020). Santos et al. (2019) developed a framework whose components are based on Lora and ARIMA (Autoregressive Integrated Moving Average) technology to enhance agricultural productivity and predict difficulties.

c. The WiMax Worldwide Interoperability data transfer range for microwave access is 1.5 Mb/s to 1 Gb/s. But presently, advanced technology has enabled high-speed data transmission (Dziyauddin et al., 2016). This communication technology may be used for wired or wireless broadband multi-access connectivity.

d. Mobile Communications Standards: The Mobile Communication Standards have evolved through 2G, 3G, and 4G. These standards are used by IoT devices to communicate over cellular networks. Farmers may monitor crop production, soil, and climate conditions, as well as discover temporal variations throughout their fields, by utilising mobile communication (Feng et al., 2019).

e. Low-rate wireless personal area networks (LR-WPANs) specify high-level communication protocols, like the specifications for ZigBee. The transfer rate for data varies from 40 Kb/s to 250 Kb/s in this communication. It offers low speed and inexpensive communication (Ray, 2018). The majority of indoor agricultural applications for LR-WPAN are on small farms and home gardens.

f. RFID collects information by giving each thing a special number that is assigned to it exclusively. RFID is made up of tags, hosts, and readers. Because tags broadcast and receive radio waves, they are also referred to as responders. RFID tags come in a variety of sizes and shapes and are divided into active and passive tags. Because passive tags are less expensive than active tags, they have more advantages than active tags. In addition to a unique ID number, tags provide environmental data on humidity, temperature, and moisture content. To identify an object, these tags are affixed to and implanted in it.
g. The ZigBee Alliance developed IEEE 802 standards, among which ZigBee is at the top. It is a collection of guidelines for low-power data-rate device-to-device networks. Faster and lower-power technology is required as a result of technological advancements and growing throughput demands. The more established technologies that offer faster data transport meet these standards. IoT sensors pick up data in an agricultural setting and send it to a distant server. Following sensing, data is gathered and analysed to aid in decision-making (Cheng and Deng, 2009).
h. MQTT is an IoT messaging protocol mainly designed for distant communications. It is a low-power-consumption protocol that saves bandwidth. Agricultural smart systems and analysis often use MQTT. Using MQTT to monitor, examine, and analyse agricultural data, gather information from field ambient, and enhance environmental conditions has resulted in the presentation of a web-based, low-cost IoT solution (Jaishetty, 2016). A low-cost irrigation system based on MQTT was established to receive and deliver sensor-collected data (Kumar and Behera, 2015).
i. According to Piti et al. (2017), SigFox is a low-data-rate, ultra-narrowband wireless cellular network that is suitable for communications between machines and the IoT. Terrasson et al. (2016) used the SigFox network to build a geographic location system that allows animals to be located in their pasture. The system suggested by Llaria et al. (2015) aids farmers in determining the location of their animals and boosts output.
j. The finest mobile communication technology for short-range communication is Bluetooth, a low-power, low-range personal area network. Numerous IoT agricultural gadgets are Bluetooth-enabled. Due to its widespread use, Bluetooth is seen as a suitable technology for applications in multi-tier agriculture (Ruiz-Garcia, 2009).

### 2.3.1 Cloud storage

The most necessary and widely adopted storage solutions are cloud and traditional databases. Because they enable the ability to store and retrieve important information when needed, these storage systems are essential for a wide range of businesses operating in diverse sectors and industries. The idea of big data, which refers to enormous amounts of datasets utilised by businesses for a variety of purposes, has emerged due to the availability of these storage systems (Monica et al., 2017).

Sensors are used to collect and interpret data when it comes to cloud computing in agriculture, particularly in irrigation-based systems. Data can be processed in cloud, and users can access the data by establishing a connection with the cloud. The employment of cloud in irrigation is usually done for storing the monitored data and then retrieving it when needed (Salvi et al., 2017). Users of cloud technology have access to both free and commercial solutions for storing, evaluating, and presenting data across a range of platforms and devices. The utilisation of this technology for the purpose of storing work-related data greatly enhances overall performance effectiveness. Research and development may be one of the most important applications of the stored data, which is accessible for a variety of reasons and at different times. Many organisations in the agriculture sector are now able to store and access information that helps improve work efficiency by applying the capabilities of cloud technology (Tyagi et al., 2017).

Moreover, alerts have been produced via cloud computing by creating irrigation algorithms. By using these notifications, numerous dangers and hazards that would have otherwise occurred have been reduced. By employing these alerts, it is easier to modify work activities and take the required safety measures to lessen the likelihood of any difficulties (Sales, 2015). Numerous cloud-related programmes have been established to aid in work performance. Although each programme has its own purpose and importance, its implementation is dependent on criteria like cost, services, adaptability, and other factors. Considering the hazards, losses, and complexities involved, maintaining the irrigation system could be viewed as a complicated task. Those engaged in irrigation activities use cloud technology to lower risks and enhance work outputs in order to meet predetermined goals (Fernández-Ahumada et al., 2019).

## 2.4 EXECUTION OF FIELD OPERATIONS

The development of GPS and global navigation satellite system (GNSS) has improved the practice of precision farming. The exact position-based crop and soil data help farmers in the development of field maps (Ntinas et al., 2017). Sensor arrays used on GPS-enabled harvesters gather comparable data. This includes multidimensional visuals and real-time sensors that evaluate each detail, from plant health to water quality to chlorophyll levels (McBratney and Pringle, 1999). To allocate resources effectively, these data are combined with satellite imaging and flexible measuring technology (VRT), which includes seeds, sprays, and other materials. Recent technological developments, however, have made it possible to deploy real-time sensors that are situated directly on the ground. These sensors may wirelessly communicate data without requiring human presence (Reyns et al., 2002).

### 2.4.1 Artificial neural network

Machine learning (ML) is a subfield of artificial intelligence that has made it possible for computers to learn, analyse, and predict without explicit programming. It is usually categorised into predicted or supervised learning and unsupervised learning (Murphy, 2012). Finding "interesting patterns" in the data is the aim of unsupervised learning, where only inputs are provided (Wayayok et al., 2014 and Shah et al., 2021). Without any input labelled data, the algorithm is left to discover and examine the underlying pattern in this situation. By organising data patterns, the algorithm gains knowledge and makes predictions about the result. Clustering and association issues are a couple of instances of unsupervised learning (Shah et al., 2021).

Deep learning is a part of machine learning. Among the deep learning architectures that offer an attractive supervised learning framework are artificial deep neural networks (Goodfellow et al., 2016; Brownlee, 2021). Numerous agricultural processes, including flower species identification, plant disease prediction and detection, weed classification and detection, crop yield forecasting, and plant species identification and classification, use machine learning and deep learning algorithms (Liakos et al., 2018).

A group of artificial neurons makes up an artificial neural network (ANN). They are usually arranged in layers and connected. A layer is composed of neurons with activation functions that are coupled to one another. An input layer, one or more hidden layers, and an output layer make up a neural network. Inputs from the outside world are received by the input layer, which uses a weighted link to route them to the hidden levels. The hidden layers then pass the obtained values to the output layer, where feature extractions and calculations are performed. The hidden layers are activated using standard nonlinear activation functions such as sigmoid, softmax, tanh, and ReLU (Rectified Linear Unit). These networks are commonly known as feed-forward neural networks or multilayer perceptrons. Figure 2.2 depicts a neural network that is feed-forward.

The basic goal of training a neural network is to minimise the error in prediction of the output value by backwardly modifying the weights on every connection. Back-propagation is the term for this procedure (Chollet, 2021). Next, the back-propagation process employs a stochastic gradient descent technique to find the lowest value in the weight space. The training process is subsequently concluded by using the weights that were collected and found to be able to minimise the loss/cost function (Chollet, 2021).

### 2.4.2 Convolution neural networks

Convolutional neural networks (CNNs) are a type of deep neural network technique that is mostly used for problems linked to computer vision

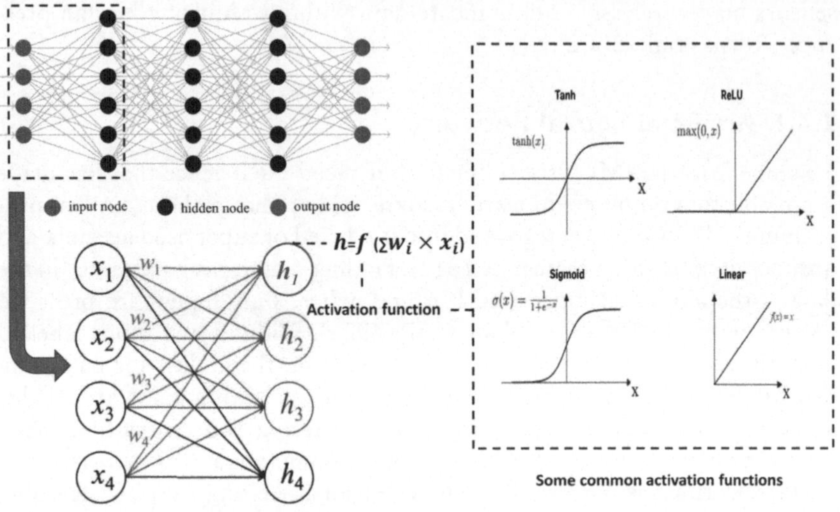


*Figure 2.2* A feed forward neural network.

(Goodfellow et al., 2016). Applications such as object detection, image classification, and instance segmentation frequently use them. CNNs are unique in that they can recognise and comprehend the temporal or spatial connection in the data. These have great success in real-world applications. Instead of using standard matrix multiplication in one of its layers, CNNs apply a unique type of mathematical operation known as a convolution operation (Goodfellow et al., 2016).

Three layers make up a convolution neural network (ConvNet): convolutional layer, the pooling layer, and the fully linked or dense layer. CNN architectures can be created by stacking or aligning all those layers in a certain order. A CNN is shown in Figure 2.3. The core structural component

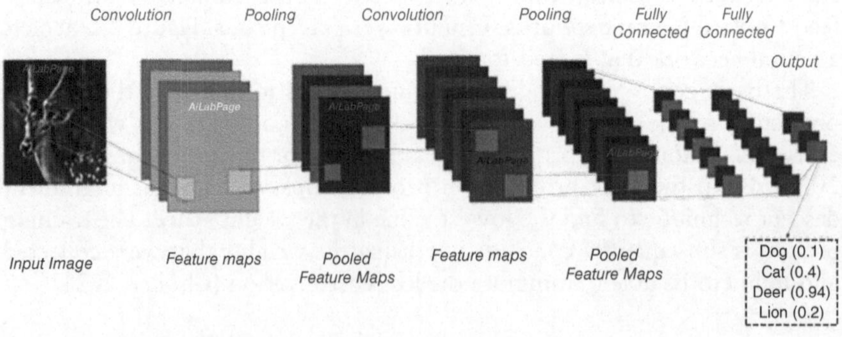


*Figure 2.3* Example showing the processing of a convolutional neural network. **Anonymous 1, 2021.**

of CNNs is the convolution layer. It is made up of kernels that independently convolve the input image to produce a set of feature maps. The three factors that regulate the activation map's size or volume are strides, depth, and zero paddings (Anonymous 1, 2021). Here, depth is stride, which indicates the number of pixels it must traverse over the image at a time, and the number of kernels used for convolution over the input image (Anonymous 1, 2021). The activation map shrinks in size, and information in the corners is lost when the kernel is convolved over the input image. The activation map's output volume can be better controlled by applying zero values at the corners, which is achieved through the use of zero padding. A nonlinear activation function is also connected to each neuron in the network to enable it to comprehend complex input. ReLU is a popular activation function because it gives the network the capacity to produce precise predictions (Anonymous 1, 2021).

The pooling layer preserves important information when the images are larger. However, computing expenses and parameters for training are reduced by primarily decreasing the spatial size. Another name for pooling is subsampling or downsampling. Each of the image's depth dimensions is pooled separately. But the pooling layer also lessens the tendency to overfit during training. In practice, among various pooling methods, max pooling with a 2×2 filter and stride = 2 is frequently utilised for superior outcomes (Anonymous 1, 2021).

### 2.4.3 Object detection methods

The most difficult aspect of picture classification problems is object recognition (also known as detection, recognition, or identification). It includes creating certain bounding boxes and giving each bounding box a class label name in order to classify different items in an image and localise the objects that are discovered (Brownlee, 2021). Another issue in computer vision is called instance or semantic segmentation, when objects are indicated with particular pixels or masks rather than a bounding box (Brownlee, 2021).

Deep learning approaches to object detection are much more successful than machine learning approaches and do not need human feature extraction. You Only Look Once (YOLO), Region-Based Convolutional Neural Network (R-CNN), and Single Shot Multi Detector (SSD) are some useful techniques proposed for object identification and localisation tasks (Girshick, R., 2015; He et al., 2020; Liu et al., 2016; Girshick et al., 2014).

R-CNN is a three-step process that was suggested in 2014. First, an image's regions that might contain objects (about 2000 proposals) are located using a selective search technique (Ren et al., 2015; Girshick, 2015). Subsequently, characteristics are extracted using a CNN, and then they are classified. This method has a slow item identification speed and requires a lot of processing. The Fast R-CNN approach was developed to address some of the issues and speed up operations (Girshick, 2015). Here, the first stage involves using a CNN to extract all of the features. Next, features for a particular input region are extracted using an ROI pooling layer, and the output is fed to a fully connected layer that divides and sends it to two classifiers that carry out bounding box regression and classification (Girshick, 2015).

There is also another approach. Faster R-CNN was developed by Ren et al., and it exceeded both the prior models in speed and detection (Girshick, 2015). They presented the regional proposal network (RPN) technique and integrated it into a single mode (Girshick, 2015). The regions are proposed using RPN, and the Fast R-CNN detector makes use of the regions that are suggested. Another technique for pixel-level semantic segmentation is mask R-CNN, which is an extension of the faster R-CNN (He et al., 2020). Along with classification and localisation, it was presented as a third branch built on the faster R-CNN architecture. It is a pixel-by-pixel prediction network that is fully coupled. It is quick, but it's not precisely calibrated for speed and precision (He et al., 2020).

A distinct approach from the R-CNN family of approaches mentioned above is used by Redmon et al. in their popular object identification tool, YOLO. Bounding boxes and class probabilities are predicted from the photos using a single neural network. Redmon et al. (2016) reported that their fast YOLO model and base model can process images in real time at 45 FPS and 155 FPS with twice mAP. It tends to generate more localisation errors, even though it was said to be quick and performed faster than the most advanced R-CNN family algorithms (Redmon et al., 2016). Another method Wei Liu et al. suggested to use a single neural network to identify objects in photos is called SSD (Liu et al., 2016). In one go, it generates region proposals and instantly recognises the objects within the suggested region. RPN-based techniques, on the other hand, have a greater mAP than either YOLO or faster R-CNN since they require two shots, making them slower than SSD (Liu et al., 2016).

### 2.4.4 Transfer learning technique

In machine learning and deep learning, transfer learning is a common technique, involving the application of knowledge gained from a pre-trained neural network model to a new model, as outlined by Goodfellow et al. (2016). Olivas et al. (2009) alternatively define it as an optimisation method facilitating rapid progress during the learning process for a distinct task, and this can be expressed mathematically.

### 2.4.5 Applications in agriculture

#### *2.4.5.1 Disease diagnosis in plants*

Computer vision, image processing, object recognition, artificial intelligence, and machine learning algorithms play a pivotal role in diagnosing and detecting plant diseases. Studies, including one by Mohanty et al. in 2016, demonstrated a high disease prediction accuracy of 99.35% across 26 diseases in 14 crop types using well-known architectures like AlexNet and GoogleNet. These findings highlight the substantial impact of deep machine learning interventions on advancing agricultural research.

#### *2.4.5.2 Yield estimation*

Enhancing contemporary farming systems involves optimising yield estimation through innovative techniques. Traditional methods, relying on past experience and seasonal weather conditions, often prove inadequate due to dynamic factors like changes in soil-water dynamics and climate variability. Machine learning algorithms, utilising time-series yield data and various spatial agricultural inputs, offer an advanced approach to improving yield prediction models. Noteworthy models encompass artificial neural networks (ANNs), support vector machines (SVM), Bayesian networks (BN), least squared support vector machines (LS-SVM), backpropagation networks (BPN), and convolutional neural networks (CNN).

Studies highlight the practical applications of these models. For instance, farmers optimised work schedules and financial plans using SVM algorithms on coffee plantations. Another study successfully predicted tomato crop yield using an ANN model and unmanned aircraft systems (UAS) for spatial and temporal remote sensing. Additionally, a study considered soil conditions, meteorological factors, and management practices from 1980 to 2015 to forecast crop yield using a CNN-RNN model in maize and soybean fields across 13 US states. Moreover, a recent study introduced a deep learning-based 3D CNN model that outperformed advanced machine learning methods in forecasting soybean crop yield.

#### *2.4.5.3 Identification of plants*

Advancements in computer vision and machine learning techniques are revolutionising plant-related research, benefitting ecologists, biologists, taxonomists, and other researchers. These technologies streamline categorisation tasks, reducing dependence on expert availability. Deep learning models, particularly those addressing image recognition, surpass traditional methods in plant leaf identification. For instance, a suggested ResNet26-based deep learning model achieved a recognition score of 91.78% using 10,000 photos from the BJFU100 dataset. In a similar vein, it demonstrated an impressive 99.65% accuracy in categorising 32 leaf architectures from the publicly available Flavia leaf dataset.

Studies reveal that plant leaf veins, along with colours and shapes, serve as input elements for leaf identification. These strategies are now being integrated into mobile applications, such as Flora-incognita, which allows users to identify plants by capturing a photo. Notably, Pl@ntNet, a mobile application, currently identifies 27,909 plant species and houses a vast database comprising 1,794,096 photos of various plants.

#### 2.4.5.4 Weed detection

Effectively managing weeds in crops is a challenging task for farmers, and inadequate control can significantly impact agricultural yields. Early detection and classification of weeds are crucial for enhancing crop productivity, given their rapid growth and competition for nutrients with crops. Machine learning algorithms, such as SVM and ANN, have demonstrated high accuracy in weed classification across various crops. Notably, a deep learning model developed by Ashqar et al., using Aarhus University's plant seedling dataset, achieved a remarkable 99.48% accuracy in classifying 12 weed types from 5000 photos.

In a separate study, Smith et al. employed CNNs and transfer learning strategies to classify grass, dock, and clover, achieving 94.9% accuracy in weed classification. Transfer learning, identified as a powerful technique by Smith et al., proves effective even with limited datasets. Sujaritha et al. designed a fuzzy real-time classifier that achieved a high accuracy level of 92.9% for weed identification in sugarcane fields. Recent developments in deep learning architectures, however, show promise for improving weed control instruments even more and investigating novel management strategies.

Real-time weed identification by robots is identified as a valuable addition to sustainable weed control systems, potentially reducing reliance on excessive pesticide use. Researchers, including Dadashzadeh et al. and Liakos et al., emphasise the significance of integrating such technologies into weed management strategies. In a recent study by Shah et al. in 2021, ANN was employed for weed management using a pre-trained neural network model. The faster R-CNN approach yielded an overall mean average precision of approximately 31%, showcasing promising results in plant and weed identification. The study also introduced a straightforward yet effective stem estimation method based on bounding box localisation for identified weeds.

## 2.5 IoT-BASED TECHNOLOGIES

The IoT is enabling smart agriculture, which is revolutionising a number of industries, including traffic management, housing, cities, healthcare, and agriculture. This technology optimises agricultural practices by connecting tools and equipment, enhancing precision in fertiliser and irrigation decisions (Kumar & Periasamy, 2021). Smart systems elevate the accuracy and efficiency of equipment for animal husbandry and plant monitoring. Wireless sensor networks (WSNs) collect data from various sensors, and coupling cloud services with IoT facilitates remote data analysis for informed decision-making (Farooq et al., 2020).

Implementing information and communication technology (ICT), automated equipment's control systems, and ground sensors like robots and self-driving cars is crucial for intelligent farm management. Satellites, high-speed internet, and advanced mobile devices play a vital role in developing smart systems. James et al. (2019) successfully utilised IoT in real time to monitor leaf diseases affecting paddy and banana crops, providing timely insights to farmers. Addressing the annual destruction of crops due to pests, diseases, and inadequate monitoring, sensors, and smart systems enable real-time monitoring of weather conditions, fertility levels, and precise fertiliser requirements.

Farooq et al. (2020) conducted a survey on IoT in agriculture, revealing significant research focus on precision agriculture, irrigation monitoring, soil monitoring, temperature, animal monitoring, humidity monitoring, air and disease monitoring, and water monitoring. Fertilisation monitoring received relatively less attention in the study publications.

The benefits of IoT-based smart agriculture extend to disease prediction, plant protection, irrigation, and product quality improvement in real time (Adamides et al., 2020). The foundational architecture for agricultural IoT systems is depicted in Figure 2.4, with two notable EU initiatives, SENSEI and IoT-A, aiming to integrate RFID, wireless sensing, and execution networks into an open, service-oriented internet structure (Presser et al., 2009; Walewski, 2011). The latter project, IoT-A, focuses on designing a reference structural model abstracted from the IoT process rather than a specific application structure.

A key component of agricultural IoT is data transmission, and WSN is a significant information transfer technology. When transmitting agricultural IoT data, communication technology should not be chosen at random; rather, it should be chosen in accordance with the unique project requirements. Some of the IoT-based research for the regulation of specific agricultural activities is presented in Table 2.2.

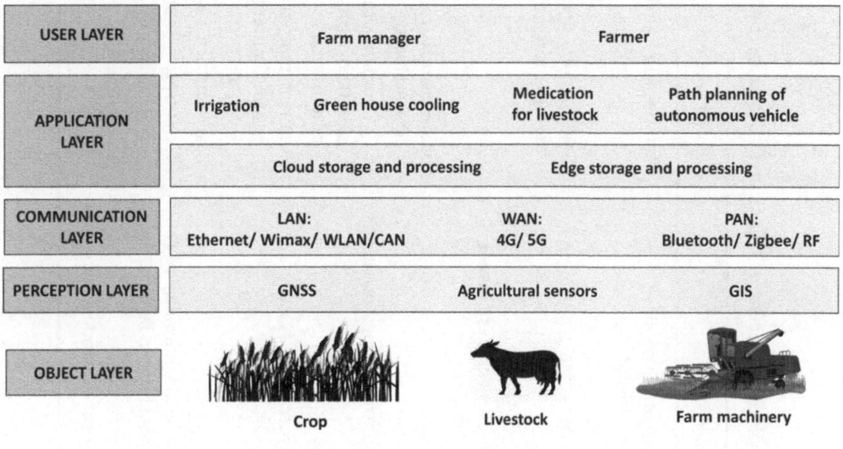


*Figure 2.4* Architecture of agricultural Internet of Things (IoT).

*Table 2.2* Internet of Things (IoT)-based agricultural applications

| *Activity* | *Parameters and sensors* | *Microcontroller* | *Data transmission* | *Reference* |
|---|---|---|---|---|
| Green house air conditioning | Temperature and humidity | PIC18F4520 microcontroller | Fuzzy controller and ZigBee wireless sensor network (WSN) | Yang et al., 2013 |
| Irrigation in green house | Soil humidity, light, air temperature, and humidity | ARM processor | WSN, transceiver hardware, and ZigBee and the CC2530 protocol stack | Xu et al., 2012 |
| Monitoring pest, insect trapper | Camera | CC1110F32 SoC | 3G/4G internet communication | López et al., 2012 |
| Irrigation | Soil moisture, temperature, and humidity | Arduino UNO | Global System for Mobile Communication (GSM) | Krishnan et al., 2020 |
| Indoor lighting conditions | $CO_2$, humidity, temperature, spectrums, photoperiod, and intensity | Artmel-ATMega | Zigbee, Dash-7, and Wi-Fi | Harun et al., 2019 |
| Hydroponic system | Ambient temperature, lighting, humidity, water level, pH | Raspberry Pi, Arduino UNO | | Mehra et al., 2018 |
| Irrigation | Soil moisture, rain, water flow | Raspberry pi | Wi-Fi | Kumar et al., 2018 |
| Irrigation, humidity of the mushroom farm | Temperature, humidity, and soil moisture | nodeMCU | Internet | Muangprathub et al., 2019 |
| Soil fertiliser recommendation system | Soil nutrient | Raspberry pi | Internet | Bodake et al., 2018 |

## 2.6 UNMANNED AERIAL VEHICLE

Precision agriculture utilises drones equipped with specialised cameras, such as multispectral or RGB cameras, to capture diverse field images. These images are processed using photogrammetric algorithms to generate orthophotos containing additional information per pixel, including near-spectra values and red values, essential for analysing indicators like normalised difference vegetation index (NDVI) maps.

Thermal cameras measure infrared radiation, detect temperature variations, aid in the early identification of diseases, assessment of irrigation efficiency, and livestock monitoring. LIDAR sensors employ laser beams to create high-resolution 3D terrain maps, providing detailed elevation data. NDVI sensors calculate a vegetation index based on NIR and red light reflection, indicating vegetation health.

GPS receivers offer precise location data for accurate navigation and georeferenced data collection. The integration of these sensors empowers agricultural drones to deliver comprehensive and precise information, facilitating informed decisions for optimal crop management and increased productivity. Figure 2.5 illustrates the array of sensors associated with UAVs for agricultural applications.

Hyperspectral cameras, a pivotal component in agricultural drone technology, surpass traditional cameras by capturing a broader range of wavelengths. This capability allows for detailed insights into crops, soil, and environmental conditions, enabling the detection of diseases, nutrient deficiencies, weed presence, various crop species, soil conditions, and stress in plants caused by factors like water scarcity, pests, or environmental conditions. Photogrammetry software is employed to construct 3D maps and models using drone-captured data.

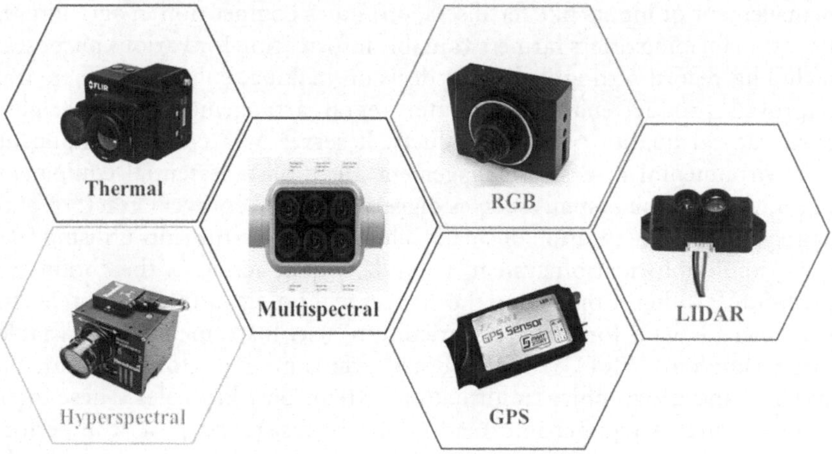


*Figure 2.5* Sensors used for agricultural applications in drones.

These drones can take pictures and provide other position-related data, such as height, which the software may use to execute algebraic map operations and produce accurate location maps. Through the employment of numerous applications at different levels, these topographic maps can be utilised to establish a connection between plant health and the environment. This connection can be made between the health of the plants and chemicals like growth regulators and herbicides, as well as water and fertilisers. Aerial and satellite photos, weather forecasts, flexible fertilisation treatments, and plant health indicators were the first wave of direct agricultural transformation. Location maps, soil data, and machine data for more precise planting were incorporated in the second wave.

Crop monitoring can be achieved by evaluating the health of the plants, determining nutrient deficits, and spotting symptoms of illness or stress. Tractor-mounted equipment is used in traditional crop spraying techniques, which can be time-consuming and lead to an uneven application of fertiliser or insecticides. Spraying system-equipped drones provide a more accurate and effective substitute. Drones fitted with thermal imaging cameras are able to keep an eye on animal conditions and behaviour. Since drones are aerial vehicles, they can manoeuvre across tough terrain with ease and reach places that would be otherwise difficult to reach. Drones can be used to seed in these kinds of areas.

## 2.7 PRECISION AGRICULTURE AND VARIABLE RATE TECHNOLOGY

Precision agriculture aims to increase field level management by comparing farming practices with plant needs (e.g., fertiliser inputs); and enhancing economic aspects through effective methods, such as improved management of inputs like fertilisers, promotes competition in agriculture. Precision farming offers farmers valuable information for various purposes, including record-setting, informed decision-making, enhanced adherence, improved product marketing, better rental arrangements, and heightened natural quality of farm products. It serves as a crucial component in environmental diversity management and utilises essential equipment like auto-direction systems such as diggers, planters, sprayers, tractors, and aggregates. The integration of small technological instruments utilising GIS (geographic information system) is fundamental, acting as the "brain" in precision farming. Connecting the correct technology and data systems to machinery is vital for effective agricultural use. Instruments like variable rate technology (VRT), remote sensors, grid sampling, global positioning systems, and geographical information systems play key roles. These tools enable farmers to gather information on soil resistance, past crop performance, and residual nitrogen in the soil by driving tractors around fields defined with GPS receivers. Utilising satellite or aerial photos, farmers create maps of their fields, providing accurate geolocation. This technology, initially applied in the late 1980s for mapping pH changes and fertiliser recommendations, has evolved with manufacturing sensors and GPS receivers, now covering vast hectares of land. Information technology, as illustrated in Figure 2.6, plays a pivotal role in boosting agricultural production and productivity.

Today, numerous studies and innovations focus on the IoT, an interconnected system of physical objects incorporating electronics for data collection, robotics, drone and satellite imagery, and smartphone applications to advance precision technology. Precision technology offers myriad benefits, including increased plant productivity and output through the precise application of inputs like water, fertiliser, and herbicides at optimal growth stages. This not only enhances yields but also enables more efficient agricultural management, leading to significant reductions in the use of fertilisers and other resources, translating to cost savings for farmers. VRT, as outlined in Table 2.3, represents some of the research in this domain.

By applying the right amount of chemicals at the correct location and time, precision technology contributes to the improvement of plants, soil, and groundwater, thereby optimising the entire agricultural cycle. Sustainable agriculture aims to ensure food security while considering social, economic, and environmental factors for long-term productivity. Variable usage of organic or pesticide levels minimises inputs, reducing costs, and mitigating

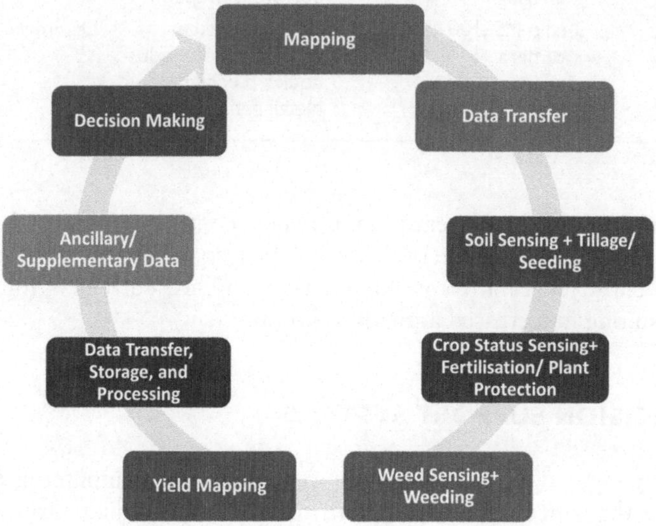


*Figure 2.6* Precision agriculture information flow in crop production.

*Table 2.3* Development of VRT for different agricultural operations

| *Operation* | *Sensor* | *Actuator* | *Reference* |
|---|---|---|---|
| Seeding | Reference data, GPS, encoder | Electric DC motor | Jafari et al. (2010) |
| Phosphorous | Portable VIS–NIR spectrophotometer sensor with a measuring range of 305–1711 nm | Electric | Maleki et al. (2008) |
| Nitrogen | N-sensors | Electric | Heib et al. (2021) |
| | Green Seeker N-sensor | Hydraulic | Mirzakhaninafchi et al. (2021) |
| Fertiliser | Prescription map based | Pneumatic actuator | Alameen et al. (2019) |
| | | Electric step motor | Song et al. (2021) |
| | | Electric step motor | Loghavi and Forouzanmehr (2010) |
| Centre pivot irrigation | RTK GPS | Solenoid valve | Li et al. (2023) |
| Spraying | Camera and image processing | Solenoid valve | Tewari et al. (2020) |
| Fungicide | 3-Chip-CCD multispectral camera | VarioSelect system with multiple nozzle bodies (Lechler, Metzingen, Germany) | Tackenberg et al. (2016) |
| Pesticide | R-G-NIR multispectral imaging | Robotic arm with six degrees of freedom | Oberti et al. (2016) |
| Herbicide | 3-chip MS2100 CCD camera | VarioSelect system with multiple nozzle bodies (Lechler, Metzingen, Germany) | Dammer (2016) |

environmental impacts. Integration of remote control systems, such as GPS, further enhances efficiency by lowering fuel consumption in agriculture. Consequently, precision technology has streamlined current farming practices, resulting in increased agricultural output.

## 2.8 DECISION SUPPORT SYSTEMS

Based on input data, decision support systems are computer-based programmes that offer recommendations to farmers regarding various activities. With these agricultural decision support systems (ADSS), the timing of irrigation, pollination, phenology, fertilisation, harvesting, and selling can all be accurately regulated to maximise earnings. These systems receive their input from a variety of sources, including weather, soil, workflow data—which includes the scheduling of various field operations—sensor data from devices, equipment data, and high-definition visual footage taken by fixed-wing, drone, and satellite aircraft.

An ADSS should offer farmers alternatives that compromise while investigating the trade-off between conflicting objectives. An informed decision must take into account information from a variety of data sources, including economic (farm income, profit, and GDP), social (public support subsidies and farm employment), input (seasonality, requirements, and available technology), environmental (temperature, humidity, rainfall, sunshine hours, soil data), and crop health (nitrogen content, disease, coverage, and water stress). The trade-off between the aforementioned factors is evaluated using a linear model based on these indicators. Farmers may immediately manage and monitor irrigation processes with a graphical user interface. ADSS uses a variety of models, including Bayesian, genetic programming, supervised learning, and probabilistic and optimisation models. Table 2.4 presents the results of some studies on the use of several models for DSS.

*Table 2.4* Methods and approaches for DSS in different agricultural applications

| *Model* | *Operation* | *Reference* |
|---|---|---|
| Embedded Markov model | Analyse production factors associated with reproductive and replacement management | Pla et al. (2004) |
| Milord | Identification of sows having low prolificacy performance | Pomar and Pomar (2005) |
| Reverse engineering argumentation | Food quality in agri-food chains | Thomopoulos et al. (2015) |
| Support vector machine | Early detection of sugar beet disease | Rumpf et al. (2010) |
| Back-propagation neural network | Spatial distribution of soybean yield | Irmak et al. (2006) |
| ANN | Strength of key spectral vegetation indices for corn crop yield prediction | Panda et al. (2010) |
| SVM and ANN | Predict nitrogen application and weed infestation rates from classification of hyper spectral image | Karimi et al. (2006) |
| Bayesian hierarchical model | Forecast combining information from multiple monthly surveys | Wang et al. (2012) |
| SVM | Predict demand for perishable farm products | Du et al. (2013) |
| Mathematical model | Prediction of production costs | Alhadidi et al. (2012) |
| Genetic algorithm | Forecasting agricultural output | Ou (2012) |

## 2.9 CONCLUSION

Continuous developments in sensors, communication, computing, machine learning techniques, and their integration to provide real-time remote access to farm data have made timely and accurate management possible. Inputs can be reduced by site-specific applications using VRT, while output can be maximised by reducing crop loss through proper monitoring. UAV, satellite, or aircraft-based data are analysed to gather information on crop health. On the basis of the collected location specific crop data, VRTs are operated. DSS helps farmers decide appropriate steps to take to maximise profit on the basis of the data fed. Automatic detection, communication, and actuation have reduced human efforts and saved precious time. The underlying technologies, like ANN, AI, object detection, etc., play a crucial role in all of the above applications. Besides these advancements, automatic systems are prone to error and challenging to operate under varying soil, environmental, and crop conditions. Dust, temperature, varying light intensity, and rainfall are some of the prime challenges in farm automation. Adaptation of these technologies is very low in developing countries due to a lack of knowledge, reliability, non-availability of maintenance services, a larger number of small and marginal farmers with fragmented land holdings, and non-levelled land separated by high bunds. Hence, scale-appropriate technologies must be developed to suit farmers in developing countries. Semi-automatic processes or products for farm mechanisation are adoptable as an intermediate step before proceeding to complete automation in developing countries.

## BIBLIOGRAPHY

Adamides, G., Kalatzis, N., Stylianou, A., Marianos, N., Chatzipapadopoulos, F., Giannakopoulou, M., Papadavid, G., Vassiliou, V., Neocleous, D., 2020. Smart Farming Techniques for Climate Change Adaptation in Cyprus. *Atmosphere* 11 (6), 557. https://doi.org/10.3390/atmos11060557

Ahmed, N., De, D., & Hussain, I. (2018a). Internet of Things (IoT) for smart precision agriculture and farming in rural areas. *IEEE Internet of Things Journal*, 5(6), 4890–4899.

Ahmad, A. A., Omar, N. R. N., & Ariff, E. E. E. (2018). The Smart Agriculture Revolution Transforming Controlled Environment Farming in Malaysia. https://ap.fftc.org.tw/article/3679

Alameen, A.A., Al-Gaadi, K.A., Tola, E., 2019. Development and performance evaluation of a control system for variable rate granular fertilizer application. *Computers and Electronics in Agriculture* 160, 31–39.

Alexandratos N, Bruinsma J. 2012. *World agriculture towards 2030/2050: the 2012 revision*. Work. Pap. 12–03, Food Agric. Organ., Rome

Alhadidi, B., Alafeef, A., Al-Hiari, H., 2012. Symbolic regression of crop pest forecasting using genetic programming. *Turkish Journal of Electrical Engineering and Computer Sciences* 20, 1332–1342.

Anand, A., de Veciana, G., Shakkottai, S., 2020. Joint Scheduling of URLLC and eMBB Traffic in 5G Wireless Networks. *IEEE/ACM Transactions on Networking* 28 (2), 477–490.

Anonymous 1 (2021). Convolutional Neural Networks (CNNs/ConvNets). Available online: https://cs231n.github.io/convolutional-networks/ (accessed on 23 January 2021).

Anonymous 3 (2017). The future of food and agriculture–Trends and challenges. *Annual Report, FAO.*

Ashapure, A., Oh, S., Marconi, T. G., Chang, A., Jung, J., Landivar, J., & Enciso, J. (2019, May). Unmanned aerial system based tomato yield estimation using machine learning. In *Autonomous air and ground sensing systems for agricultural optimization and phenotyping IV* (Vol. 11008, pp. 171–180). SPIE.

Ashqar, B. A., Abu-Nasser, B. S., & Abu-Naser, S. S. (2019). Plant seedlings classification using deep learning.

Bacco, M., Barsocchi, P., Ferro, E., Gotta, A., & Ruggeri, M. (2019). The digitisation of agriculture: a survey of research activities on smart farming. *Array*, 3, 100009.

Bodake, K., Ghate, R., Doshi, H., Jadhav, P., & Tarle, B. (2018). Soil based fertilizer recommendation system using Internet of Things. *MVP Journal of Engineering Sciences*, 1(1), 13–19.

Brownlee, J. (n.d.). What Is Deep Learning? Available online:

Brownlee, J. A. (2021). Gentle Introduction to Object Recognition with Deep Learning. Available online: https://machinelearningmastery.com/object-recognition-with-deep-learning/ (accessed on 23 January 2021).

Chen, Z., & Shah, T. M. (2019). An Introduction to the Global Soil Status. *RUVIVAL Publication Series*, 7–17.

Cheng X, Deng Z. Constructing a Large Scale Wireless Sensors Network and Resources Scheduling among Multiple Personal Area Networks. *International Journal of Distributed Sensor Networks* 2009;5(1):76–76. https://doi.org/10.1080/15501320802561690

Chollet, F. (2021). *Deep learning with Python*. Simon and Schuster.

Dadashzadeh, M., Abbaspour-Gilandeh, Y., Mesri-Gundoshmian, T., Sabzi, S., Hernández-Hernández, J. L., Hernández-Hernández, M., & Arribas, J. I. (2020). Weed classification for site-specific weed management using an automated stereo computer-vision machine-learning system in rice fields. *Plants*, 9(5), 559.

Dammer, K. H. (2016). Real-time variable-rate herbicide application for weed control in carrots. *Weed Research*, 56(3), 237–246.

Dias, J., Grilo, A. Multi-hop LoRaWAN uplink extension: specification and prototype implementation. *Journal of Ambient Intelligence and Humanized Computing* 11, 945–959 (2020). https://doi.org/10.1007/s12652-019-01207-3

Du, J. X., Wang, X. F., & Zhang, G. J. (2007). Leaf shape based plant species recognition. *Applied Mathematics and Computation*, 185(2), 883–893.

Du, X.F., Leung, S.C., Zhang, J.L., Lai, K.K., 2013. Demand forecasting of perishable farm products using support vector machine. *International Journal of Systems Science* 44, 556–567.

Dziyauddin, R.A., Doufexi, A., Kaleshi, D., Sam, S.M., & Mohamed, N. (2016). Performance Evaluation of Dynamic Burst Mapping in a WiMAX System. *Wireless Personal Communications*, 91, 1191–1212.

Farooq, A., Jia, X., Hu, J., & Zhou, J. (2019, September). Knowledge transfer via convolution neural networks for multi-resolution lawn weed classification.

In *2019 10th Workshop on Hyperspectral Imaging and Signal Processing: Evolution in Remote Sensing (WHISPERS)* (pp. 01–05). IEEE.

Farooq, M.S., Riaz, S., Abid, A., Umer, T., Zikria, Y.B., 2020. Role of IoT technology in agriculture: A systematic literature review. *Electronics* 9 (2), 319.

Feng, X., Yan, F. & Liu, X. Study of Wireless Communication Technologies on Internet of Things for Precision Agriculture. *Wireless Personal Communications* 108, 1785–1802 (2019). https://doi.org/10.1007/s11277-019-06496-7

Fernández-Ahumada, L. M., Ramírez-Faz, J., Torres-Romero, M., & López-Luque, R. (2019). Proposal for the design of monitoring and operating irrigation networks based on IoT, cloud computing and free hardware technologies. *Sensors*, 19(10), 2318.

Filippi, P., Jones, E. J., Wimalathunge, N. S., Somarathna, P. D., Pozza, L. E., Ugbaje, S. U., ... & Bishop, T. F. (2019). An approach to forecast grain crop yield using multi-layered, multi-farm data sets and machine learning. *Precision Agriculture*, 20, 1015–1029.

Fuentes, A., Yoon, S., Kim, S. C., & Park, D. S. (2017). A robust deep-learning-based detector for real-time tomato plant diseases and pests recognition. *Sensors*, 17(9), 2022.

Gagliardi, G., Lupia, M., Cario, G., Cicchello Gaccio, F., D'Angelo, V., Cosma, A. I. M., & Casavola, A. (2021). An internet of things solution for smart agriculture. *Agronomy*, 11(11), 2140.

Girshick, R. (2015). Fast r-cnn. arXiv. arXiv *preprint arXiv:1504.08083*.

Girshick, R., Donahue, J., Darrell, T., & Malik, J. (2014). Rich feature hierarchies for accurate object detection and semantic segmentation. In *Proceedings of the IEEE conference on computer vision and pattern recognition* (pp. 580–587).

Goëau, H., Bonnet, P., Joly, A., Bakić, V., Barbe, J., Yahiaoui, I., ... & Péronnet, A. (2013, October). Pl@ ntnet mobile app. In *Proceedings of the 21st ACM international conference on Multimedia* (pp. 423–424).

Goodfellow, I., Bengio, Y., & Courville, A. (2016). *Deep learning*. MIT press.

Grinblat, G. L., Uzal, L. C., Larese, M. G., & Granitto, P. M. (2016). Deep learning for plant identification using vein morphological patterns. *Computers and Electronics in Agriculture*, 127, 418–424.

Harun, A. N., Mohamed, N., Ahmad, R., & Ani, N. N. (2019). Improved Internet of Things (IoT) monitoring system for growth optimization of Brassica chinensis. *Computers and Electronics in Agriculture*, 164, 104836.

He, Z. B., Jiang, Y., Zhang, W. J., Zapien, J. A., Liu, Y., Zhang, W. F., & Bello, I. (2009). Growth, evolution and photocatalytic activity of ZnO nano back-tapered arrays. *Physica Status Solidi (a)*, 206(1), 94–100.

Heib, A., Paraforos, D.S., Sharipov, G.M., Griepentrog, H.W., 2021. Modeling and simulation of a multi-parametric fuzzy expert system for variable rate nitrogen application. *Computers and Electronics in Agriculture* 182, 106008.

Hemathilake, D. M. K. S., & Gunathilake, D. M. C. C. (2022). High-productive agricultural technologies to fulfill future food demands: Hydroponics, aquaponics, and precision/smart agriculture. In *Future Foods* (pp. 555–567). Academic Press. https://doi.org/10.1109/TNET.9010.1109/TNET.2020.2968373. https://machinelearningmastery.com/what-is-deep-learning/ (accessed on 23 January 2021).

Irmak, A., Jones, J., Batchelor, W., Irmak, S., Boote, K., Paz, J., 2006. Artificial neural network model as a data analysis tool in precision farming. *Transactions of the ASABE* 49, 2027–2037.

Jafari, M., Hemmat, A., Sadeghi, M., 2010. Development and performance assessment of a DC electric variable-rate controller for use on grain drills. *Computers and Electronics in Agriculture* 73, 56–65.

Jaishetty, S.A. (2016). IoT sensor network based approach for agricultural field monitoring and control. *International Journal of Research in Engineering and Technology*, 05, 45–48.

James, A., Saji, A. Nair, Joseph, D., 2019. CropSense–A smart agricultural system using IoT. *Journal of Electronic Design Engineering* 5 (3), 1–7.

K. He, G. Gkioxari, P. Dollár and R. Girshick, "Mask R-CNN," *IEEE Transactions on Pattern Analysis and Machine Intelligence*, vol. 42, no. 2, pp. 386–397, 1 Feb. 2020, https://doi.org/10.1109/TPAMI.2018.2844175

Karimi, Y., Prasher, S., Patel, R., Kim, S., 2006. Application of support vector machine technology for weed and nitrogen stress detection in corn. *Computers and Electronics in Agriculture* 51, 99–109.

Khaki, S., Wang, L., & Archontoulis, S. V. (2020). A cnn-rnn framework for crop yield prediction. *Frontiers in Plant Science*, 10, 1750.

Kopittke, P. M., Menzies, N. W., Wang, P., McKenna, B. A., & Lombi, E. (2019). Soil and the intensification of agriculture for global food security. *Environment International*, 132, 105078.

Krishnan, R. S., Julie, E. G., Robinson, Y. H., Raja, S., Kumar, R., & Thong, P. H. (2020). Fuzzy logic based smart irrigation system using internet of things. *Journal of Cleaner Production*, 252, 119902.

Kumar, S. and P. Behera, "A low cost smart irrigation control system," *2015 2nd International Conference on Electronics and Communication Systems (ICECS)*, Coimbatore, India, 2015, pp. 1146–1152, https://doi.org/10.1109/ECS.2015.7124763

Kumar, A., Surendra, H., Mohan, K. Muthu Valliappan, N. Kirthika. Internet of things based smart irrigation using regression algorithm. *2017 Int. Conf. Intell. Comput. Instrum. Control Technol. ICICICT 2017*. 2018-Janua (2018), pp. 1652–1657, https://doi.org/10.1109/ICICICT1.2017.8342819

Kumar, T. U., Periasamy, A., 2021. IoT Based Smart Farming (E-FARM)'S. *International Journal of Recent Advances in Multidisciplinary Topics* 2 (4), 85–87.

Lai, Y. J., Kuo, W. H., Chiu, W. T. *et al.* Accelerometer-assisted 802.11 rate adaptation on mobile WiFi access. *Journal on Wireless Communications and Networking* 2012, 246 (2012). https://doi.org/10.1186/1687-1499-2012-246

Li, M., Wang, Y., Guo, H., Ding, F., & Yan, H. (2023). Evaluation of variable rate irrigation management in forage crops: Saving water and increasing water productivity. *Agricultural Water Management*, 275, 108020.

Liakos, K. G., Busato, P., Moshou, D., Pearson, S., & Bochtis, D. (2018). Machine learning in agriculture: A review. *Sensors*, 18(8), 2674.

Liebman, M., Baraibar, B., Buckley, Y., Childs, D., Christensen, S., Cousens, R., ... & Riemens, M. (2016). Ecologically sustainable weed management: how do we get from proof-of-concept to adoption?. *Ecological Applications*, 26(5), 1352–1369.

Liu, W., Anguelov, D., Erhan, D., Szegedy, C., Reed, S., Fu, C. Y., & Berg, A. C. (2016). Ssd: Single shot multibox detector. In *Computer Vision–ECCV 2016: 14th European Conference, Amsterdam, The Netherlands, October 11–14, 2016, Proceedings, Part I 14* (pp. 21–37). Springer International Publishing.

Llaria, A., Terrasson, G., Arregui, H., & Hacala, A. (2015, March). Geolocation and monitoring platform for extensive farming in mountain pastures. In *2015 IEEE International Conference on Industrial Technology (ICIT)* (pp. 2420–2425). IEEE.

Loghavi M, Forouzanmehr E. Design, Development and Field Evaluation of a Map-Based Variable Rate Granular Application Control System. In *Proceedings of the 2010 Pittsburgh, Pennsylvania*, June 20–June 23, 2010; 2010: 1.

López, O., Rach, M. M., Migallon, H., Malumbres, M. P., Bonastre, A., & Serrano, J. J. (2012). Monitoring pest insect traps by means of low-power image sensor technologies. *Sensors*, 12(11), 15801–15819.

Maleki, M.R., Mouazen, A.M., De Ketelaere, B., Ramon, H., De Baerdemaeker, J., 2008. On-the-go variable-rate phosphorus fertilisation based on a visible and nearinfrared soil sensor. *Biosystems Engineering* 99, 35–46.

Marios, S., & Georgiou, J. (2017, October). Precision agriculture: Challenges in sensors and electronics for real-time soil and plant monitoring. In *2017 IEEE Biomedical Circuits and Systems Conference (BioCAS)* (pp. 1–4). IEEE.

McBratney, A.B., Pringle, M.J. Estimating Average and Proportional Variograms of Soil Properties and Their Potential Use in Precision Agriculture. *Precision Agriculture* **1**, 125–152 (1999). https://doi.org/10.1023/A:1009995404447

Mehra, M., Saxena, S., Sankaranarayanan, S., Tom, R. J., & Veeramanikandan, M. (2018). IoT based hydroponics system using Deep Neural Networks. *Computers and Electronics in Agriculture*, 155, 473–486.

Mirzakhaninafchi, H., Singh, M., Bector, V., Gupta, O., Singh, R., 2021. Design and development of a variable rate applicator for real-time application of fertilizer. *Sustainability* 13, 8694.

Mohanty, S. P., Hughes, D. P., & Salathé, M. (2016). Using deep learning for image-based plant disease detection. *Frontiers in Plant Science*, 7, 1419.

Monica, M., Yeshika, B., Abhishek, G. S., Sanjay, H. A., & Dasiga, S. (2017, October). IoT based control and automation of smart irrigation system: An automated irrigation system using sensors, GSM, Bluetooth and cloud technology. In *2017 International Conference on recent innovations in signal processing and embedded systems (RISE)* (pp. 601–607). IEEE.

Muangprathub, J., Boonnam, N., Kajornkasirat, S., Lekbangpong, N., Wanichsombat, A., & Nillaor, P. (2019). IoT and agriculture data analysis for smart farm. *Computers and Electronics in Agriculture*, 156, 467–474.

Murphy, K. P. (2012). *Machine learning: a probabilistic perspective*. MIT press.

Ahmed, N., De, D. and Hussain, I., "Internet of Things (IoT) for Smart Precision Agriculture and Farming in Rural Areas," *IEEE Internet of Things Journal*, vol. 5, no. 6, pp. 4890–4899, Dec. 2018b, https://doi.org/10.1109/JIOT.2018.2879579

Nowak, B. Precision Agriculture: Where do We Stand? A Review of the Adoption of Precision Agriculture Technologies on Field Crops Farms in Developed Countries. *Agricultural Research* **10**, 515–522 (2021). https://doi.org/10.1007/s40003-021-00539-x

Ntinas, G. K., Neumair, M., Tsadilas, C. D., & Meyer, J. (2017). Carbon footprint and cumulative energy demand of greenhouse and open-field tomato cultivation systems under Southern and Central European climatic conditions. *Journal of Cleaner Production*, 142, 3617–3626.

Oberti, R., Marchi, M., Tirelli, P., Calcante, A., Iriti, M., Tona, E., … & Ulbrich, H. (2016). Selective spraying of grapevines for disease control using a modular agricultural robot. *Biosystems Engineering*, 146, 203–215.

Ofori-Dwumfuo, G.O., & Salakpi, S.V. (2011). WiFi and WiMAX Deployment at the Ghana Ministry of Food and Agriculture.

Olivas, E. S., Guerrero, J. D. M., Martinez-Sober, M., Magdalena-Benedito, J. R., & Serrano, L. (Eds.). (2009). *Handbook of research on machine learning applications and trends: Algorithms, methods, and techniques: Algorithms, methods, and techniques*. IGI global.

Ou, S.-L., 2012. Forecasting agricultural output with an improved grey forecasting model based on the genetic algorithm. *Computers and Electronics in Agriculture* 85, 33–39.

Palombi, L., & Sessa, R. (2013). Climate-smart agriculture: sourcebook. *Food and Agriculture Organization of the United Nations (FAO)*.

Pan, S. J., & Yang, Q. (2009). A survey on transfer learning. *IEEE Transactions on Knowledge and Data Engineering*, 22(10), 1345–1359.

Panda, S.S., Ames, D.P., Panigrahi, S., 2010. Application of vegetation indices for agricultural crop yield prediction using neural network techniques. *Remote Sensing* 2, 673–696.

Piti, A., Verticale, G., Rottondi, C.E., Capone, A., & Schiavo, L.L. (2017). The Role of Smart Meters in Enabling Real-Time Energy Services for Households: The Italian Case. *Energies*, 10, 1–25.

Pla, L., Pomar, C., Pomar, J., 2004. A sow herd decision support system based on an embedded Markov model. *Computers and Electronics in Agriculture* 45, 51–69.

Pomar, J., Pomar, C., 2005. A knowledge-based decision support system to improve sow farm productivity. *Expert Systems with Applications* 29, 33–40.

Presser, M., Barnaghi, P. M., Eurich, M., & Villalonga, C. (2009). The SENSEI project: Integrating the physical world with the digital world of the network of the future. *IEEE Communications Magazine*, 47(4), 1–4.

Ramirez, W. A., Mishra, G., Panda, B. K., Jung, H. W., Lee, S. H., Lee, I., & Singh, C. B. (2022). Multispectral camera system design for replacement of hyperspectral cameras for detection of aflatoxin B1. *Computers and Electronics in Agriculture*, 198, 107078.

Ratnaparkhi, S., Khan, S., Arya, C., Khapre, S., Singh, P., Diwakar, M., & Shankar, A. (2020). Withdrawn: Smart agriculture sensors in IOT: A review.

Raun, W. R., Solie, J. B., Johnson, G. V., Stone, M. L., Lukina, E. V., Thomason, W. E., & Schepers, J. S. (2001). In-season prediction of potential grain yield in winter wheat using canopy reflectance. *Agronomy Journal*, 93(1), 131–138.

Ray, P. P. (2018). A survey on Internet of Things architectures. *Journal of King Saud University, Computer and Information Sciences*, 30(3), 291–319.

Redmon, J., Divvala, S., Girshick, R., & Farhadi, A. (2016). You only look once: Unified, real-time object detection. In *Proceedings of the IEEE conference on computer vision and pattern recognition* (pp. 779–788).

Ren, S., He, K., Girshick, R., & Sun, J. (2015). Faster r-cnn: Towards real-time object detection with region proposal networks. arXiv 2015. *arXiv preprint arXiv:1506.01497*.

Reyns, P., Missotten, B., Ramon, H., & De Baerdemaeker, J. (2002). A review of combine sensors for precision farming. *Precision Agriculture*, 3, 169–182.

Roopaei, M., Rad, P., & Choo, K. K. R. (2017). Cloud of things in smart agriculture: Intelligent irrigation monitoring by thermal imaging. *IEEE Cloud Computing*, 4(1), 10–15.

Ruiz-Garcia, L., Lunadei, L., Barreiro, P., & Robla, I. (2009). A review of wireless sensor technologies and applications in agriculture and food industry: state of the art and current trends. *Sensors*, 9(6), 4728–4750.

Rumpf, T., Mahlein, A.-K., Steiner, U., Oerke, E.-C., Dehne, H.-W., Plümer, L., 2010. Early detection and classification of plant diseases with support vector machines based on hyperspectral reflectance. *Computers and Electronics in Agriculture* 74, 91–99.

Russello, H. (2018). Convolutional neural networks for crop yield prediction using satellite images. *IBM Center for Advanced Studies*.

Saleem, M. H., Potgieter, J., & Arif, K. M. (2019). Plant disease detection and classification by deep learning. *Plants*, 8(11), 468.

Sales, N., Remédios, O., & Arsenio, A. (2015, December). Wireless sensor and actuator system for smart irrigation on the cloud. In *2015 IEEE 2nd World Forum on internet of things (WF-IoT)* (pp. 693–698). IEEE.

Salvi, S., Jain, S. F., Sanjay, H. A., Harshita, T. K., Farhana, M., Jain, N., & Suhas, M. V. (2017, February). Cloud based data analysis and monitoring of smart multi-level irrigation system using IoT. In *2017 International Conference on I-SMAC (IoT in Social, Mobile, Analytics and Cloud) (I-SMAC)* (pp. 752–757). IEEE.

Santos, U.J., Pessin, G., Costa, C.A., & Righi, R.D. (2019). AgriPrediction: A proactive internet of things model to anticipate problems and improve production in agricultural crops. *Computers and Electronics in Agriculture*, 161, 202–213.

Shah, T. M., Nasika, D. P. B., & Otterpohl, R. (2021). Plant and weed identifier robot as an agroecological tool using artificial neural networks for image identification. *Agriculture*, 11(3), 222.

Shapira, U., Herrmann, I., Karnieli, A., & Bonfil, J. D. (2010). Weeds detection by ground-level hyperspectral data. *Theory Into Practice*, 38, 27–33.

Sheng, P., Guo, Y., & Li, P. (2012). Intelligent measurement and control system of facility agriculture based on ZigBee and 3G. *Nongye Jixie Xuebao= Transactions of the Chinese Society for Agricultural Machinery*, 43(12), 229–233.

Smith, L. N., Byrne, A., Hansen, M. F., Zhang, W., & Smith, M. L. (2019, September). Weed classification in grasslands using convolutional neural networks. In *Applications of Machine Learning* (Vol. 11139, pp. 334–344). SPIE.

Song, C., Zhou, Z., Zang, Y., Zhao, L., Yang, W., Luo, X., Jiang, R., Ming, R., Zang, Y., Zi, L., 2021. Variable-rate control system for UAV-based granular fertilizer spreader. *Computers and Electronics in Agriculture* 180, 105832.

Sujaritha, M., Annadurai, S., Satheeshkumar, J., Sharan, S. K., & Mahesh, L. (2017). Weed detecting robot in sugarcane fields using fuzzy real time classifier. *Computers and Electronics in Agriculture*, 134, 160–171.

Sun, Y., Liu, Y., Wang, G., & Zhang, H. (2017). Deep learning for plant identification in natural environment. *Computational Intelligence and Neuroscience*, 2017.

Sutton (2018). 5G network architecture. *Journal of The Institute of Telecommunications Professionals* 12, 8–15, https://doi.org/10.1016/b978-0-08-102267-2.00001-4

Tackenberg, M., Volkmar, C., & Dammer, K. H. (2016). Sensor-based variable-rate fungicide application in winter wheat. *Pest Management Science*, 72(10), 1888–1896.

Tang, Yu, Dananjayan, Sathian, Hou, Chaojun, Guo, Qiwei, Luo, Shaoming, He, Yong, 2021. A survey on the 5G network and its impact on agriculture: Challenges and opportunities. *Computers and Electronics in Agriculture* 180, 105895. https://doi.org/10.1016/j.compag.2020.105895

Terrasson, G., Llaria, A., Marra, A., & Voaden, S. (2016, July). Accelerometer based solution for precision livestock farming: geolocation enhancement and

animal activity identification. In *IOP Conference Series: Materials Science and Engineering* (Vol. 138, No. 1, p. 012004). IOP Publishing.
Tewari, V. K., Pareek, C. M., Lal, G., Dhruw, L. K., & Singh, N. (2020). Image processing based real-time variable-rate chemical spraying system for disease control in paddy crop. *Artificial Intelligence in Agriculture*, 4, 21–30.
Thomopoulos, R., Croitoru, M., Tamani, N., 2015. Decision support for agri-food chains: a reverse engineering argumentation-based approach. *Ecological Informatics* 26, 182–191.
Tilman D, Balzer C, Hill J, Befort BL. 2011. Global food demand and the sustainable intensification of agriculture. *PNAS* 108:20260–20264
Tyagi, S., Obaidat, M. S., Tanwar, S., Kumar, N., & Lal, M. (2017, December). Sensor cloud based measurement to management system for precise irrigation. In *GLOBECOM 2017-2017 IEEE Global Communications Conference* (pp. 1–6). IEEE.
Wäldchen, J., & Mäder, P. (2018). Machine learning for image based species identification. *Methods in Ecology and Evolution*, 9(11), 2216–2225.
Walewski, J. W. (2011). Internet-of-Things Architecture IoT-A Project Deliverable D1. *2-Initial Architectural Reference Model for IoT*.
Wang, J.C., Holan, S.H., Nandram, B., Barboza, W., Toto, C., Anderson, E., 2012. A Bayesian approach to estimating agricultural yield based on multiple repeated surveys. *Journal of Agricultural, Biological, and Environmental Statistics* 17, 84–106.
Wayayok, A., Soom, M. A. M., Abdan, K., & Mohammed, U. (2014). Impact of mulch on weed infestation in system of rice intensification (SRI) farming. *Agriculture and Agricultural Science Procedia*, 2, 353–360.
Williams, J., Tadesse, A., Sam, T., Sun, H., & Montañez, G. D. (2020). Limits of transfer learning. In *Machine Learning, Optimization, and Data Science: 6th International Conference, LOD 2020, Siena, Italy, July 19–23, 2020, Revised Selected Papers, Part II 6* (pp. 382–393). Springer International Publishing.
Wu, S. G., Bao, F. S., Xu, E. Y., Wang, Y. X., Chang, Y. F., & Xiang, Q. L. (2007, December). A leaf recognition algorithm for plant classification using probabilistic neural network. In *2007 IEEE international symposium on signal processing and information technology* (pp. 11–16). IEEE.
Xu, J., Gu, B., & Tian, G. (2022). Review of agricultural IoT technology. *Artificial Intelligence in Agriculture*, 6, 10–22.
Xu, Z., Lou, B., & Shao, G. (2012). An Intelligent Irrigation System for Greenhouse Jonquil Based on ZigBee Wireless Sensor Networks. In *Internet of Things: International Workshop, IOT 2012, Changsha, China, August 17-19, 2012. Proceedings* (pp. 200–207). Springer Berlin Heidelberg.
Yang, M. T., Chen, C. C., & Kuo, Y. L. (2013). Implementation of intelligent air conditioner for fine agriculture. *Energy and Buildings*, 60, 364–371.
Zhao, W. (2017, August). Research on the deep learning of the small sample data based on transfer learning. In *AIP Conference Proceedings* (Vol. 1864, No. 1). AIP Publishing.

Chapter 3

# Smart solutions and sustainability for agriculture using deep learning

*Dev Batra, Jaspreet Kaur, Chirag Tyagi, Chaitanya Sharma and Divyanshu Joshi*

## 3.1 INTRODUCTION

In the realm of agriculture, the health of plants plays a pivotal role in ensuring food security and sustaining ecosystems. However, the constant threat of plant diseases poses a significant challenge to global crop production. Farming has been around for a long time, but now there are new challenges. More people need more food, and factors like climate change make it harder to grow crops. The old ways of checking for diseases by just looking at plants are not fast enough or good enough anymore. We need new ideas!

The urgency to address this issue has driven the integration of cutting-edge technologies, with deep learning emerging as a revolutionary tool in the field of plant pathology. Imagine if computers could learn and understand things like humans do. That is what deep learning is all about. These smart computers use their 'brains' to recognize patterns and make decisions. It is like having a super-smart assistant helping farmers spot plant diseases quickly and accurately.

Plants get sick from all sorts of things, like tiny bugs or bad weather. Figuring out what's wrong can be like solving a puzzle because different plants show different signs of being sick. Deep learning can understand these signs better than humans and can work for all kinds of crops, anywhere in the world.

For deep learning to do its magic, it needs lots of information – pictures of healthy and sick plants, details about different diseases, and more. This data is like the superpower that makes the computer so good at its job. We will explore how scientists collect and use this data to train the computer to be a plant disease detective.

We will see how deep learning, a subset of artificial intelligence, has demonstrated unparalleled capabilities in image recognition, pattern analysis, and decision-making. In recent years, its application in plant disease detection has gained momentum, offering a promising solution to enhance the efficiency and accuracy of diagnosis. This book chapter explores the evolution and potential impact of deep learning techniques in the realm of plant

 DOI: 10.1201/9781003628392-3

disease detection, shedding light on the transformative journey from traditional methods to state-of-the-art technology.

## 3.2 HISTORICAL CONTEXT OF PLANT DISEASE DETECTION

A long time ago, when people first started farming, they noticed that sometimes their plants got sick, but they did not really understand why. They could see that some plants were not growing well, but the reasons behind these issues remained a mystery. This early period marked the beginning of observations regarding plant diseases, where farmers noticed patterns without having a clear scientific understanding.

As time passed and agriculture became more organized, especially in the 19th century, scientists started studying plant diseases more seriously. They began to differentiate between diseases caused by living organisms, like tiny germs, and those caused by other factors. This period saw the formalization of plant pathology as a scientific discipline, setting the stage for a more systematic exploration of the causes and solutions to plant diseases.

Advancements in technology, particularly the improvement in microscopes, allowed scientists in the late 19th century to study pathogens at a microscopic level. This was a crucial step forward because it enabled a deeper understanding of the tiny organisms that could make plants sick. Around the same time, the germ theory of disease gained acceptance, explaining how specific microorganisms could cause diseases in plants.

The early 20th century witnessed the introduction of chemical pesticides and biological control methods, providing practical ways to manage and mitigate plant diseases. These methods marked significant progress in the fight against agricultural pests and diseases.

However, it was not until the latter half of the 20th century that molecular techniques, such as DNA analysis, became available. This technological leap allowed scientists to identify and understand plant pathogens at a more detailed level, contributing to more precise and targeted approaches in disease management.

## 3.3 CHALLENGES FACED BY TRADITIONAL METHODS

Traditional methods of plant disease detection faced challenges due to the subjective nature of human observation. Different people might interpret and identify symptoms differently, leading to inconsistencies in the assessment of plant health. This subjectivity introduced a degree of uncertainty, making it difficult to ensure accurate and standardized diagnoses across different observers. Also, examining each plant manually was a slow and labour-intensive process, especially in the context of large-scale farming operations. The delay in identifying diseases meant that by the time issues were recognized, they had often spread to other plants, making it challenging to contain and manage the impact effectively. Quick and timely detection is crucial for preventing widespread crop damage.

Another significant drawback of traditional methods was their tendency to miss the early signs of plant diseases. Identifying diseases in their initial stages is essential for effective intervention and control. When these early signs were overlooked, diseases could progress, making it harder to implement timely and targeted measures to mitigate their impact. With the expansion of modern agriculture, the sheer volume of data generated by farms became overwhelming for traditional methods. It was akin to having numerous puzzle pieces without the capability to assemble them into a coherent picture of overall plant health. The inability to handle large-scale data hindered the efficiency of traditional approaches in providing comprehensive insights into the health of crops.

Traditional methods sometimes also struggled with precision in distinguishing between different diseases and often misidentified stress factors affecting plants. This lack of accuracy hindered the implementation of targeted solutions, as misdiagnoses could lead to ineffective or inappropriate interventions. The need for more precise and reliable identification methods became evident to enhance the overall efficacy of disease management strategies.

The good update is that advancements in technology, particularly in artificial intelligence and machine learning, are addressing these challenges. Smart technologies are now being used to identify plant diseases, providing farmers quickly and accurately with better tools to protect their crops and improve overall agricultural productivity.

## 3.4 FUNDAMENTALS OF DEEP LEARNING

In the context of plant disease detection, deep learning serves as a transformative approach to analysing images and identifying diseases with a high level of accuracy. Deep learning operates on the principles of neural networks, which are computational models inspired by the human brain. These networks, particularly convolutional neural networks (CNNs) for image-related tasks, are capable of automatically learning and extracting meaningful features from plant images. The process of deep learning involves training a neural network on a vast dataset of labelled images, where each image is associated with information about whether a plant is healthy or diseased. During training, the network learns to recognize patterns and features that distinguish healthy plants from those affected by diseases as shown in Figure 3.1. This is achieved through the adjustment of internal parameters, known as weights, based on the network's performance on the training data.

Key terms in deep learning for plant disease detection include feature extraction, where the neural network identifies important characteristics in the images, such as discolourations, lesions, or patterns associated with specific diseases. As shown in Figure 3.2, feature extraction is a critical step as it allows the model to focus on relevant aspects of the plant images. Training is the phase where the neural network learns from the labelled examples, adjusting its internal parameters to improve its ability to accurately classify plants as healthy or diseased. The trained model is then validated or tested on new, unseen images to ensure its generalization and effectiveness in detecting diseases beyond the training dataset. This testing and classification methodology is briefly depicted in Figure 3.3.

As we explore the fundamentals of deep learning in the context of plant pathology, understanding terms like feature extraction, training, and validation becomes crucial for appreciating how this technology advances the field of plant disease detection.

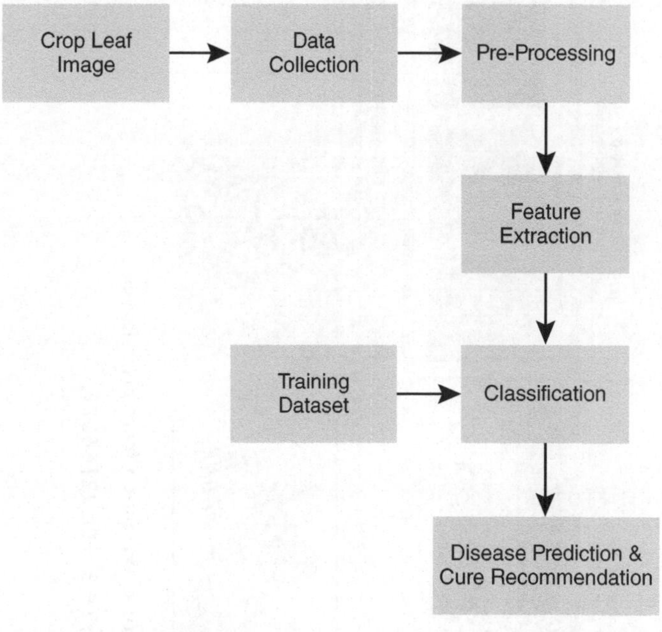


*Figure 3.1* Training system architecture.

**Source: International Journal of Creative Research Thoughts, IJCRT2305797, ISSN: 2320-2882.**

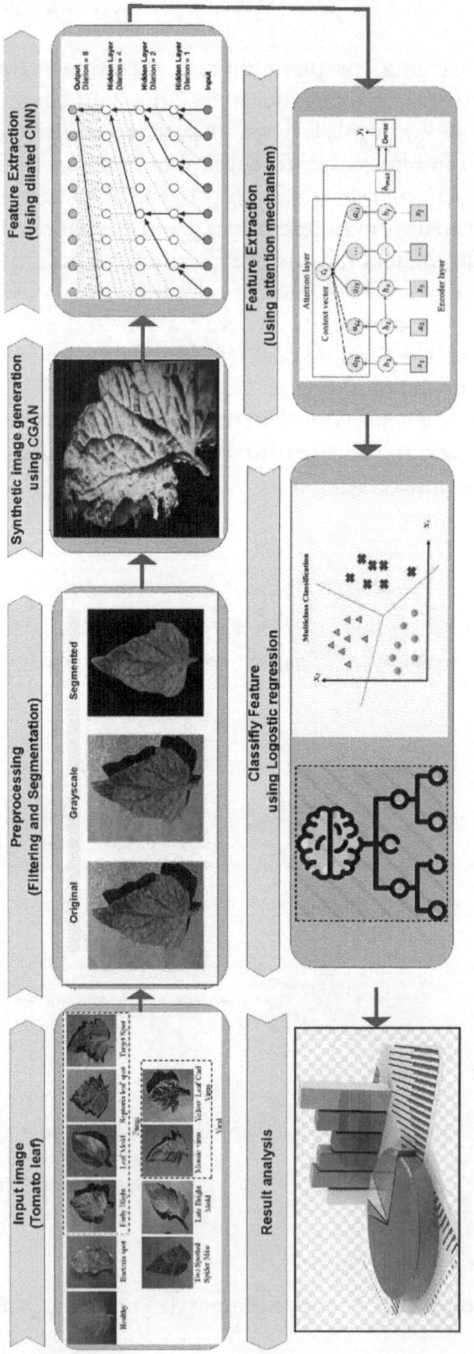


*Figure* 3.2 Feature extraction from plant leaves.

**Source: DOI: 10.3390/s22166079.**

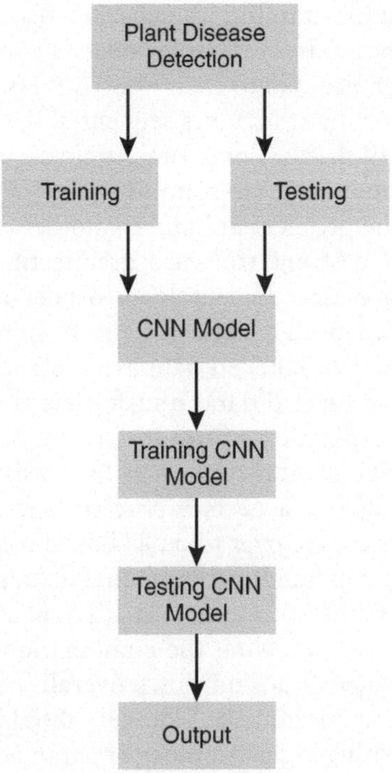


*Figure 3.3* Classification system architecture.

**Source: International Journal of Creative Research Thoughts, IJCRT-2305797, ISSN: 2320-2882.**

## 3.5 OBJECTIVE OF RESEARCH

Agriculture, crucial for meeting global food demands, underpins the economy and sustains human nutrition through grain and vegetable cultivation. However, remote farmers often suffer substantial losses due to manual monitoring and limited disease detection knowledge. Digital farming methods offer a transformative solution to swiftly detect plant leaf diseases."Employing deep learning in disease recognition minimizes bias in spot selection, fosters objective disease feature extraction, and expedites technological advancements. This chapter introduces a novel deep learning-based approach for leaf disease detection, contributing to efficient methods that benefit farmers and boost global food security.

## 3.6 DEEP LEARNING ARCHITECTURES FOR PLANT DISEASE DETECTION

Selecting the appropriate deep learning architecture for plant disease detection involves a thoughtful consideration of various factors to ensure optimal model performance. Understanding the characteristics of your dataset is a fundamental starting point. If your data primarily consists of images, CNNs are often a preferred choice due to their effectiveness in image analysis. On the other hand, if your dataset involves sequential data, such as information about the progression of diseases over time, architectures like recurrent neural networks (RNNs) may be more suitable.

The complexity of the problem at hand is another critical aspect to evaluate. For simpler tasks, a straightforward architecture might suffice, while more intricate scenarios may benefit from deeper architectures or those incorporating attention mechanisms to capture nuanced image patterns. Additionally, the amount of labelled data available for training plays a significant role. In cases of limited data, transfer learning becomes a valuable strategy, allowing the utilization of pre-trained models on diverse datasets for adaptation to specific plant disease detection tasks.

Consideration of temporal aspects is crucial, especially if the focus is on capturing how diseases evolve over time. RNNs or long short-term memory networks (LSTMs) excel in handling sequential data, providing insights into the dynamic nature of plant diseases across different stages of plant development. Ensemble methods, involving the combination of different architectures, can be advantageous to improve overall performance. Resource constraints, including computational power, should not be overlooked. Deeper architectures with a higher number of parameters require more computational resources for training. In situations where resources are limited, simpler architectures or transfer learning approaches may be more practical.

Rigorous validation and testing on separate datasets are essential to assess the generalization capabilities of different architectures. This ensures that the chosen architecture performs well not only on the training data but also on new, unseen examples. Furthermore, consulting existing literature and studies related to plant disease detection provides valuable insights and guidance, allowing all of us to learn from the experiences and successes of others in similar domains. Table 3.1 provides a concise overview of the literature survey conducted on plant disease detection in recent years.

In summary, the selection of a deep learning architecture for plant disease detection is a nuanced process, necessitating a comprehensive evaluation of data characteristics, problem complexity, available labelled data, and validation of model performance. Experimentation with different architectures, combined with a thorough understanding of the specific task requirements, is essential for making informed decisions in this evolving field.

## 3.7 BASIC TRANSFER LEARNING IMPLEMENTATION USING RESNET-50

A transfer learning-based approach is a machine learning technique that involves using knowledge gained from solving one problem and applying

*Table 3.1* Brief Overview of Literature Survey

| *Reference (Year)* | *Objectives* | *Key Findings* | *Limitations* |
|---|---|---|---|
| [1] (2022) | For precise identification of plant diseases amidst intricate natural circumstances. | YOLOV5 network outperformed both the original YOLOV5 and the YOLOX networks in terms of performance. | Detection accuracy still needs to be improved. Inconsistent dataset. |
| [2] (2021) | To evaluate the precision of pre-existing models such as VGG, ResNet, and DenseNet. | The DenseNet model achieved the highest precision of 98%. | The main problem faced is the vanishing gradient problem. |
| [3] (2022) | To diminish the parameter, count within the novel CNN through inception and residual connections. | Guaranteed high accuracies (99% approximately) on plant village, rice, and cassava dataset. | Limited classification of dataset. Untested on diverse datasets. |
| [4] (2020) | To accomplish disease identification and classify its type swiftly and precisely. | Overall precision of the classification is found to be 89% approximately. | Distinguishing is difficult in the segmentation phase because of variations. Limitations due to insufficient subclasses in data. |
| [5] (2020) | To utilize CNN concepts to identify diseases affecting paddy crops. | Efficient monitoring and control of remote farms can be achieved using just a smartphone. | Model lacks validation in global rural settings. Farmers need device to access the proposed AI. |
| [6] (2019) | To detect rice plant disease using deep learning and CNN. | Highest accuracy achieved is 91% approximately. | Unbalanced number of images Limited classification of dataset (619 images only). |

*(Continued)*

*Table 3.1* (Continued)

| *Reference (Year)* | *Objectives* | *Key Findings* | *Limitations* |
| --- | --- | --- | --- |
| [7] (2020) | Olive leaf diseases identification using CNN. | Various spot diseases like Aculus olearias can be identified on the leaf with high accuracy (95% approximately).<br>Aculus olearius and olive peacock spot diseases can be identified at a high accuracy rating | The model is expensive to implement.<br>Limited dataset diversity. |
| [8] (2023) | To detect disease using region-based convolutional neural network and ResNet. | The ResNet model demonstrated exception performance by delivering an outstanding accuracy rate of 99,2 %. | The paper does not touch upon the ethical and social implications of the implementation of such technology in agriculture.<br>Does not discuss the adaptability of the project with various agricultural scenarios. |
| [9] (2022) | To detect diseases in leaves using Deep Learning. | Exact detection of leaf diseases in low computational efforts. Using this approach, 97% accuracy is achieved. | No real time-processing capabilities of the proposed system discussed.<br>Paper does not explicitly discuss the scalability of the product. |

it to a different but related problem. In the context of deep learning, transfer learning has proven to be particularly effective, especially when dealing with limited labelled data for a specific task. At its core, ResNet introduces the concept of residual blocks, fundamentally altering the way information flows through the network.

As shown in Figure 3.4, the journey through a ResNet begins at the input layer, where the raw data is fed into the network. This could be images, text, or any other form of structured or unstructured data. From there, the data traverses through convolutional and pooling layers, forming the initial stages responsible for learning intricate hierarchical features from the input. These layers play a crucial role in extracting meaningful patterns and representations from the data.

The distinctive element of ResNet lies within its residual blocks. Each block comprises a sequence of layers, typically starting with a convolutional layer followed by a rectified linear unit (RELU) activation function. What sets ResNet apart is the inclusion of a shortcut connection, or a skip connection, allowing the network to bypass some layers. This shortcut facilitates the direct flow of information, enabling the network to focus on learning residual mappings, which are the differences between the expected output and the actual output from a specific block. Following these residual blocks, there might be additional pooling layers. These layers help reduce the dimensionality of the learned features, aiding in computational efficiency and creating more abstract representations of the data. After the stack of residual blocks and potential pooling layers, the network often includes fully connected layers. These layers serve as the pathway for synthesizing the learned representations into higher-level features, employing RELU activations to introduce non-linearity and capture complex relationships within the data.

Finally, the network culminates at the output layer, where the processed data, after traversing through the hierarchical layers, yields the desired predictions or outcomes based on the nature of the task—be it classification, regression, or another form of data-driven inference. Overall, the architecture outlined presents a comprehensive process for diagnosing plant diseases

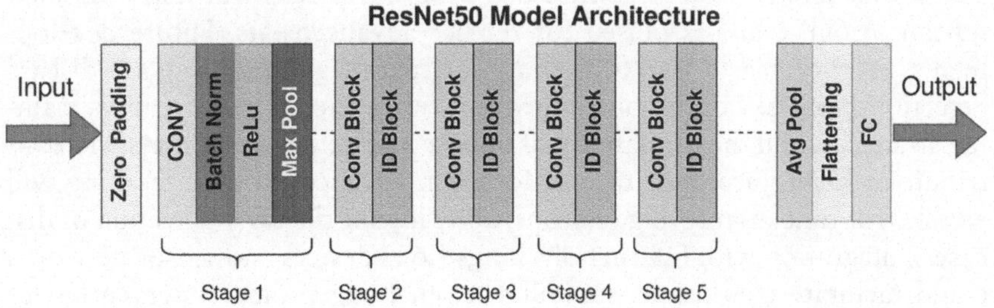


*Figure 3.4* ResNet-50 architecture.

**Source: Towards Data Science by Suvaditya Mukherjee.**

using transfer learning techniques. It begins with choosing a pre-trained deep learning model that has excelled in learning features from a large dataset. Popular choices include models like ResNet-50, VGG, or Inception, which are trained on general image recognition tasks, such as ImageNet. The general model is then modified to adapt the chosen pre-trained model to the task of plant disease detection. This involves changing the original output layer designed for the pre-training task to suit the number of classes relevant to plant diseases. Essentially, the model is reshaped to understand and identify patterns indicative of different plant diseases. Then some or all the layers are frozen in the pre-trained model. Freezing prevents these layers from being further adjusted during training, allowing the model to retain knowledge gained from the original dataset. This is especially useful when the dataset for plant diseases is small. The next step involves training the modified model on a dataset containing labelled images of healthy and diseased plants. The model learns to recognize features and patterns associated with specific diseases during this training phase.

Following diagnosis, the system proceeds to evaluate the model's performance on a separate validation set to assess its ability to generalize to new, unseen examples. Fine-tuning is done, if necessary, which involves adjusting the model parameters, potentially unfreezing some layers, and further training to improve performance on the specific plant disease detection task. This entire process is also highlighted via the diagram shown in Figure 3.5. Thus, by employing transfer learning in plant disease detection, the model benefits from the wealth of information learned during its pre-training on a diverse dataset. This accelerates the learning process and enhances the model's ability to recognize subtle visual cues indicative of various plant diseases. The approach not only addresses the challenge of limited labelled data in plant pathology but also contributes to more accurate and efficient disease identification, ultimately aiding farmers and researchers in maintaining crop health.

## 3.8 FUTURE SCOPE

The future scope of plant disease detection using deep learning holds significant promise and is poised for further advancements. Future developments are expected to lead to even more accurate and precise plant disease detection models. Continuous improvements in model architectures, training strategies, and the availability of diverse and extensive datasets will contribute to higher accuracy rates. Moreover, advances in deep learning will enable real-time monitoring of crops, allowing for the early detection of diseases. Integration with Internet of Things (IoT) devices and sensor networks could facilitate continuous surveillance, enabling prompt intervention to mitigate the spread of diseases and minimize crop losses. The deployment of deep learning models on edge devices, close to the source of data generation (e.g., drones or field sensors), will become more prevalent. This facilitates

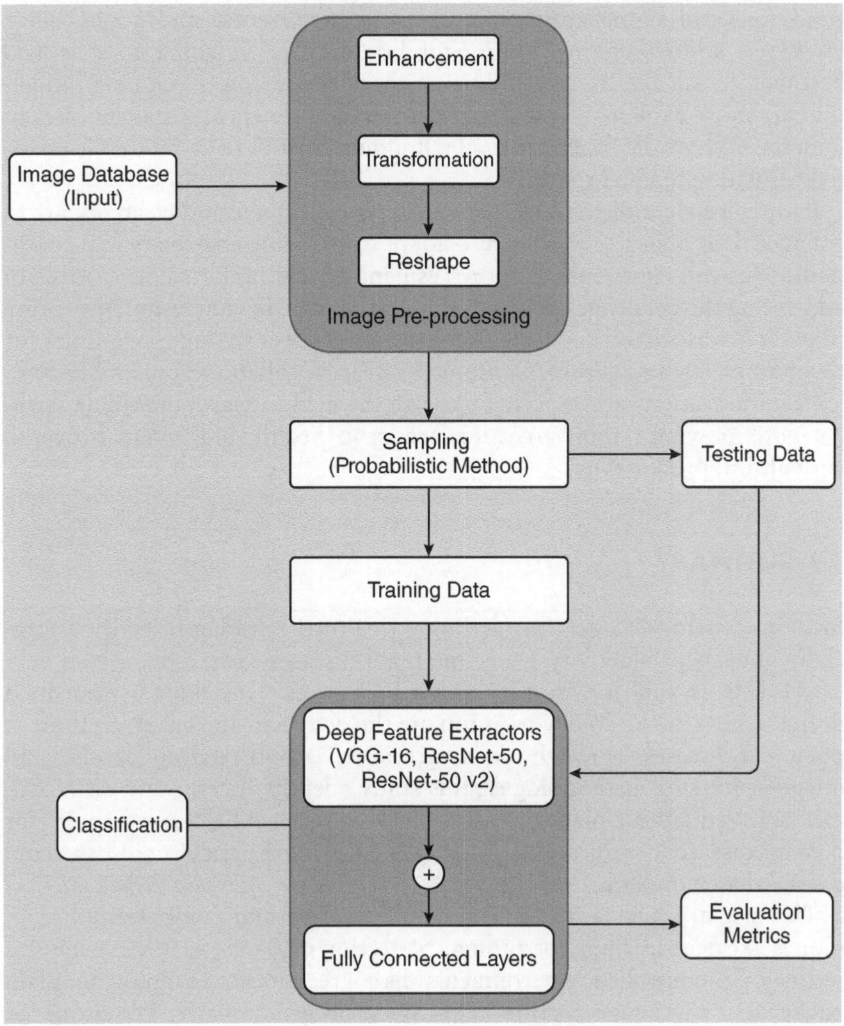


*Figure 3.5* Transfer learning implementation using ResNet-50.

**Source: Singh, A., Kaur, H. (2022). Comparative Study on Identification and Classification of Plant Diseases with the Support of Transfer Learning. In: Khanna, A., Gupta, D., Bhattacharyya, S., Hassanien, A.E., Anand, S., Jaiswal, A. (eds) *International Conference on Innovative Computing and Communications. Advances in Intelligent Systems and Computing*, vol 1387. Springer, Singapore. DOI: 10.1007/978-981-16-2594-7_31.**

on-site analysis without relying heavily on centralized computing resources, making the technology more accessible in remote agricultural areas.

Future developments will also focus on creating specialized models for different types of crops and geographical regions. Tailoring deep learning models to the unique characteristics of specific crops and environmental conditions can significantly enhance their effectiveness and applicability. However, addressing the interpretability of deep learning models will become crucial. Future advancements should focus on developing models that are more explainable, making it easier for farmers and stakeholders to understand how decisions are made. Building trust in these models is essential for widespread adoption.

Future deep learning models may incorporate climate and weather data to enhance their ability to predict and adapt to changing environmental conditions. This can contribute to more resilient agricultural systems capable of addressing the challenges posed by climate change. In conclusion, the future of plant disease detection using deep learning holds exciting possibilities for revolutionizing agriculture. Continued research, collaboration, and technological innovations are expected to drive the field forward, providing farmers with powerful tools to safeguard crop health and enhance overall agricultural productivity.

## 3.9 SUMMARY

In India, where 70% are engaged in agriculture, selecting insecticides crucially impacts productivity. Detecting plant diseases early, especially in delicate leaves, is vital for crop inception to harvest. This chapter presents a deep learning methodology to automate the detection and categorization of plant leaf diseases. Through the integration of advanced deep learning and image processing methodologies, this study adeptly discerns and differentiates between diverse plant diseases. This advancement offers prospects for prompt disease identification, potentially mitigating agricultural losses and disease dissemination.

The primary aim is the utilization of modern and computerized techniques like deep learning for anomaly detection in plants, whether in natural settings or controlled environments like greenhouses. Employing plain backgrounds in image capture helps avoid occlusion issues. The proposed method showcases remarkable accuracy in disease identification, surpassing any traditional method of plant disease detection. Notably, its strength lies in efficiently handling extensive datasets and extracting high-level features from images.

Deep learning serves not just as an analytical tool but also as a trigger for proactive responses by suggesting suitable remedies to address the diagnosed disease. Additionally, leveraging machine learning capabilities, the system can predict potential future illnesses the plant might develop. This prediction is made by analysing historical data patterns and trends to foresee potential risks or vulnerabilities in the plant's health. It underscores the importance of ethics, improved methodologies, and a steadfast commitment to always ensuring crop quality.

## REFERENCES

[1] Zhaoyi Chen, Ruhui Wu, Yiyan Lin, Chuyu Li, Siyu Chen, Zhineng Yuan, Shiwei Chen, and Xiangjun Zou, “Plant disease recognition model based on improved YOLOv5“, *Grand Challenges of Advanced Technologies in Sustainable Agriculture* 4.0, DOI: 2022, 10.3390/agronomy12020365

[2] K.P. Akshai and J. Anitha, “Plant disease classification using deep learning,” *2021 3rd International Conference on Signal Processing and Communication (ICPSC)*, 2021, pp. 407–411, DOI: 10.1109/ICSPC51351.2021.9451

[3] S.K. Mahmudul Hassan, and A. K. Maji, “Plant disease identification using a novel convolutional neural network,” *IEEE Access*, vol. 10, pp. 5390–5401, 2022, DOI: 10.1109/ACCESS.2022.3141371

[4] N. Manohar and K. J. Gowda, “Image Processing System based Identification and Classification of Leaf Disease: A Case Study on Paddy Leaf,” *2020 International Conference on Electronics and Sustainable Communication Systems (ICESC)*, 2020, pp. 451–457, DOI: 10.1109/ICESC48915.2020.9155607

[5] P. A. Harsha, S. A. Vardini, and Y. S. Devi, “Efficient Disease Detection of Paddy Crop using CNN,” *2020 International Conference on Smart Technologies in Computing, Electrical and Electronics (ICSTCEE)*, 2020, pp. 116–119, DOI: 10.1109/ICSTCEE49637.2020.9276775

[6] V. K. Shrivastava, M. K. Pradhan, S. Minz, and M. P. Thakur: Rice plant disease classification using transfer learning of deep convolution neural network. *Int. Arch. Photogramm. Remote Sens. Spat. Inf. Sci. - ISPRS Arch.*, 42 (2019) 631–635.

[7] S. Uğuz and N. Uysal: Classification of olive leaf diseases using deep convolutional neural networks. *Neural Computing and Applications*, 5 (2020) 1–17.

[8] V. Sushma Sri, V. Hima Sailu, U. Pradeepthi, P. Manogyna Sai, and D.M. Kavitha. Disease detection using region-based convolutional neural network and ResNet. *Data and Metadata*. 2023; 2:135. DOI: 10.56294/dm2023135

[9] Akshay Tike, Jambuvant Kadam, Dhairysheel Jadhav, Rushikesh Nalbalwar, A. P. Kale, Leaves disease detection using deep learning *International Journal of Advanced Research in Science, Communication and Technology* 2022, DOI: 10.48175/IJARSCT-3893

Chapter 4

# Bi-LSTM-based deep learning model for crop yield prediction

*Ajay Kumar, Kakoli Banerjee, Pradeep Kumar and Sunil Kumar Chawla*

## 4.1 INTRODUCTION

The Indian economy is based on agriculture, which provides food for both people and animals. As the world population is projected to reach 9.7 billion people by 2050, the demand for agricultural goods is forecast to rise considerably. To fulfill the rising demand for food, this calls for improved agricultural yield and the effective use of acreage. However, several challenges loom over the agricultural sector, including the impact of the COVID-19 pandemic and the consequences of climate change, which have led to adverse weather conditions and subsequent crop failures. Farmers bear the brunt of these challenges as they face the risk of crop destruction due to factors such as floods, soil degradation, climate variability, dwindling groundwater levels, and other related issues. Predicting crop yield accurately becomes exceedingly complex due to the interplay of various factors. Weather patterns, soil quality, geographical variations, pest infestations, water availability and quality, crop varieties, harvesting timings, and numerous other elements significantly influence crop output [1]. Therefore, it is imperative to consider these influencing factors when developing strategies for improved crop selection, especially considering seasonal variations.

The primary objective of agricultural yield estimation is to enhance crop production by employing various established methodologies. Machine learning (ML) has gained prominence worldwide due to its efficacy in diverse domains, including pattern recognition, defect detection, forecasting, and other applications. Leveraging ML techniques can help mitigate losses incurred from unfavorable conditions by increasing the rate of agricultural yield production. Several ML-based methods for calculating agricultural production have been studied recently. Among the techniques are artificial neural networks, decision trees, association rule mining, and multivariate regression employed in this domain. One notable characteristic of ML models is their ability to capture complex, nonlinear relationships through implicit function-based approaches to output prediction [2].

In conclusion, agriculture in India faces numerous challenges due to factors such as population growth, climate change, and the recent COVID-19


DOI: 10.1201/9781003628392-4

pandemic. Accurate estimation of crop yield becomes crucial to ensure food security and optimize agricultural productivity. By leveraging ML techniques [3] and considering various influencing factors, agricultural yield estimation models can contribute to improved crop selection strategies and enhanced production rates [4]. This study contributes to the existing knowledge by evaluating the pros and cons of different ML-based techniques, shedding light on their effectiveness in agricultural yield estimation [5].

This study aims to investigate both the advantages and disadvantages associated with different ML techniques applied in agriculture. The literature survey conducted in Section 4.2 provides an overview of existing works in this field. Section 4.3 delves into the crop yield estimation models proposed by the study [6]. In Section 4.4, the experimental findings are presented, along with the outcomes derived from the conducted experiments. Finally, Section 4.5 concludes the study by summarizing the key findings and their implications for agricultural yield estimation.

## 4.2 RELATED WORK

There has been much research done on predicting agricultural productivity; Table 4.1 gives an overview of these studies, their methodology, and their main objectives.

The study "A CNN-RNN Framework for Crop Yield Prediction" investigates the application of artificial intelligence techniques to forecast agricultural production. In prior studies, techniques including random forest (RF), multi-level linear regression (MLR), and artificial neural networks were employed. The research recommends a deep learning architecture that mixes CNNs and RNNs to improve agricultural production prediction [20]. The use of deep neural networks, CNNs, and RNNs in recent deep learning applications for agricultural production prediction is also covered. The survey emphasizes the value of precise yield prediction, the difficulties involved, and the possibility of deep learning techniques to overcome these difficulties.

An innovative deep neural network (DNN) technique for crop yield prediction utilizing historical yields and greenhouse environmental factors is presented in the literature review "Deep Learning Based Prediction on Greenhouse Crop Yield Combined TCN and RNN" [21]. The proposed method combines temporal convolutional networks (TCNs) and recurrent neural networks (RNNs) to extract representative features of the input sequence. The normalization of input data, the use of RNNs with long short-term memory (LSTM) units, and the use of TCNs for feature extraction at various temporal scales are all covered in the survey. In order to make the final forecast, a completely linked layer is used. According to experimental findings, employing RNNs and TCNs together surpasses using either network alone to estimate greenhouse crop production.

*Table 4.1* Related work on various approaches of crop yield prediction

| *S. no* | *Year* | *Objective* | *Advantage* | *Disadvantage* |
|---|---|---|---|---|
| 1 | 2020 | In this study, CNNs and RNNs are used to develop a deep learning framework for agricultural yield prediction. The CNN-RNN model performs better than other tested approaches because it can generalize to untested situations and capture temporal dependencies. The causes causing yield fluctuation can be identified using the model [7]. | For increased prediction accuracy, a hybrid CNN-RNN model combines the best features of both CNN and RNN models. The RNN model makes use of long short-term memory (LSTM) cells to better capture long- term relationships in the data. Even without access to meteorological or soil data, the model can forecast crop production using previous trends. | To successfully train the model, a lot of data is needed. The model's utilization of several layers and cells might increase its computational costs.<br>The caliber and comprehensiveness of the input data may have an impact on the model's accuracy. |
| 2 | 2020 | The objective is to develop a crop yield prediction model using multi-spectral satellite imagery and deep learning techniques [8]. | The proposed CNN-LSTM-12 model performs significantly better than previous approaches that use handcrafted features, with an improvement of over 70% in crop yield prediction. | The CNN-LSTM-12 model under-predicts the yield marginally for a majority of tehsils across all states, and for large tehsils, it consistently under estimates the yield. |
| 3 | 2020 | LSTM and bidirectional long short-term memory (bi-LSTM) models are used to map the rice fields in West Rio Grande do Sul, Brazil, utilizing Sentinel-1 time series data. In order to create an accurate strategy for crop monitoring, their performance is contrasted with that of traditional ML techniques [9]. | The use of LSTM and bi- LSTM models in crop mapping provides high accuracy, leveraging the temporal information from Sentinel-1 time series data, enabling effective monitoring and management of Rio Grande do Sul's western rice fields. | Implementing deep learning models like LSTM and bi-LSTM may require significant computational resources and expertise, limiting their accessibility and applicability in resource-constrained environments. |

| | | | | |
|---|---|---|---|---|
| 4 | 2020 | Use ML to increase agricultural production accuracy estimation, enhancing decision-making and agricultural productivity [10]. | ML enables accurate crop yield estimation, aiding decision-making and improving productivity. | Implementation challenges include data availability, computational resources, and technical expertise, limiting widespread adoption in resource-constrained agricultural settings. |
| 5 | 2021 | The aim of this project is to combine modern networks for analyzing temporal sequences to create a novel method for accurately forecasting crop yields in greenhouses, and to assess the method's efficacy using various datasets from actual greenhouses used for tomato cultivation. The objective is to give farmers and greenhouse growers a tool to help them make financially and managerially sound decisions [11]. | The suggested approach combines RNN and TCN networks to estimate greenhouse crop yields with accuracy. Compared to conventional ML and deep learning techniques, it produces predictions that are more accurate. It emphasizes how crucial historical yield data is for forecasting agricultural yields in the future. | The suggested approach needs to be tested using more datasets gathered from other farmers and places.<br>For more precise and reliable crop production prediction, a biophysical model should be used in combination with more complex network architecture. |
| 6 | 2021 | In order to enhance socioeconomic development and food security, this endeavor intends to provide a learning-based paradigm for predicting crop yields and assessing critical harvesting features. The proposed model integrates LSTM-RNN with Q-learning reinforcement to compute yield and improve accuracy [12]. | Forecasting agricultural production with accuracy to improve food security and spur economic expansion. Separating significant harvesting properties using ML methods.<br>The accuracy of yield forecast will rise with the inclusion of parametric parameters. | Only permits the use of historical data. Relies on the accuracy and reliability of data intake. Confined to the habitat and crop types that were used in the dataset. |
| 7 | 2021 | The purpose of this study is to develop a deep learning-based RNN-LSTM model for forecasting wheat crop output in northern India and compare its performance with three ML models using a benchmark dataset of 43 years [13]. | The suggested RNN-LSTM model, which is based on deep learning, provides more accurate predictions of wheat crop yield than traditional ML models. | The model may require large amounts of data for training and tuning, and may be computationally expensive. It may also be difficult to interpret the internal workings of the model. |

*(Continued)*

*Table 4.1.* (Continued)

| *S. no* | *Year* | *Objective* | *Advantage* | *Disadvantage* |
|---|---|---|---|---|
| 8 | 2021 | In order to improve agricultural production prediction, test the effectiveness of several recurrent neural network (RNN) variations for forecasting wheat crop output in a specific area of Punjab, India [14]. | Advanced ML methods like RNNs enable accurate crop yield prediction, aiding in efficient source allocation, improved planning, and increased agricultural productivity. | Implementing and maintaining sophisticated predictive models may require substantial computational resources, technical expertise, and data availability, posing challenges for widespread adoption in resource-constrained agricultural regions. |
| 9 | 2021 | This study compares the performance of recurrent and CNNs in forecasting agricultural production using environmental data and management strategies. They make use of data from 13different US states between2016 and 2018. They found that CNN-RNN model achieved better accuracy among all [15]. | The developed model offered a method for predicting crop yields based on deep learning that correctly projected corn and soybean yields over the entire US Corn Belt. | A black box that is included in many ML approaches was incorporated into the produced model. |
| 10 | 2022 | Enhance Indian agriculture through precision techniques, considering variables like farming season, soil type, and location to provide farmers more control, prevent farming abandonment, and promote unsustainable practices. | Precision agriculture enables farmers to make informed decisions based on specific variables, optimizing crop choices, reducing waste, increasing productivity, and promoting unsustainable practices. | Implementation challenges such as the initial cost of technology adoption, limited access to resources in rural areas, and the need for technical knowledge may hinder widespread adoption and benefit for all farmers. |
| 11 | 2022 | Compare the accuracy and training times of the gated recurrent unit (GRU) and LSTM algorithms for estimating wheat production [8]. | GRU and LSTM techniques enable accurate wheat production evaluation, aiding in agricultural planning decision-making. | The LSTM approach may be less effective in time-sensitive applications because of its lengthier training period than the GRU method. |

| | | | | |
|---|---|---|---|---|
| 12 | 2022 | Improve agricultural productivity by using LSTM-based ML to estimate crop yield based on meteorological data. This will also give farmers useful information for better decision-making[16]. | LSTM-based ML enables accurate crop yield prediction, optimizing agricultural production and resource allocation. | Implementing LSTM models may require computational resources, expertise, and limited accessibility in resource-constrained agricultural settings. |
| 13 | 2022 | To correctly and successfully monitor rice production in subtropical hilly areas for sustainable agriculture and global food system reform, ahigh-precision rice mapping technique utilizing SAR data and the DB-bi-LSTM model must be developed [17]. | The proposed method enables high-precision rice mapping, promoting sustainable agriculture and contributing to global food system reform through accurate monitoring and information on rice cultivation. | Implementing the proposed method may require access to synthetic aperture radar (SAR) data, computational resources, and technical expertise, limiting its widespread adoption and applicability in resource-constrained areas. |
| 14 | 2022 | Develop a bi-LSTM-based classification model using Sentinel data to identify regional crop types for agricultural monitoring [18]. | The bi-LSTM-based classification model enables the accurate identification of regional crop types using Sentinel data, facilitating the transition to remote sensing-based agricultural monitoring under the new Common Agricultural Policy (CAP2020). | The model's general acceptance and usefulness in resource-limited places may be hindered by the requirements for its implementation, which might include computational resources, technological know-how, and access to Sentinel data. |
| 15 | 2022 | Based on climatological and soil data, the Smart Crop Selection (SCS) method uses ML. The accuracy of the SCS model varies between 97% and 98%. The ability to predict 11 crops has been taught [19]. | The ML algorithms used in the proposed research provide more accuracy at a lower processing cost than those used in earlier studies. | Other factors and crops might be added to this system. You might also look at ML techniques like CNN and LSTM, which are more precise and efficient. |
| 16 | 2023 | Create a model for predicting future soil moisture using RNN-LSTM and meteorological information to ascertain irrigation requirements for crop development and to improve irrigation techniques based on variations in soil moisture caused by weather [5]. | The created RNN-LSTM soil moisture prediction model allows for accurate irrigation scheduling, maximizing water use in agriculture and fostering sustainable crop development under fluctuating weather circumstances. | Implementing the model may require sensors for soil moisture and temperature measurements, access to weather data, and technical expertise in data analysis, which could pose challenges for resource-constrained farmers. |

The literature review for this study emphasizes how crucial crop production forecasting throughout the growing season is to India's agricultural industry. It highlights the drawbacks of human surveys and introduces the use of satellite data and deep learning for yield prediction. The suggested approach beats current approaches by a considerable margin since it operates directly on unprocessed satellite photos and takes contextual information into account [22]. The experimental findings show how well the strategy predicts crop production at the tehsil level. This investigation demonstrates how deep learning and satellite imaging have the potential to enhance agricultural production estimation in the setting of India.

The literature review offered covers the investigation of temporal backscattering signals in remote sensing data as well as the use of image denoising algorithms. The study assesses how well several filtering techniques including the Sentinel time series data have less noise thanks to 3D Gamma and S-G filtering. The results demonstrate that these filters work well together, with the S-G filter further enhancing the temporal profiles without sacrificing data integrity and the 3D Gamma filter minimizing speckles. Using backscattering values, the technique also investigates the temporal fingerprints of other land cover types, including rice paddy fields, riparian woodlands, and grasslands. RNN models' classification performance for determining land use and land cover is also contrasted with that of traditional ML methods [23]. Compared to standard ML techniques, the RNN models, notably LSTM and bidirectional long short-term memory (bi-LSTM), exhibit superior accuracy and statistical benefits in the study's key classes while also being more computationally efficient. Overall, the research demonstrates the benefits of RNN models in remote sensing applications, the efficacy of picture denoising methods, and the identification of temporal signatures [24].

Precipitation, relative humidity, and soil temperature that were shown to be the most important variables in the study of the RNN-LSTM model for forecasting soil moisture in soybean production were significant factors influencing soil moisture content. Accuracy varied with soil depth, with the highest accuracy achieved at 20 cm depth. The validation performance was better in SET II compared to SET I, indicating the influence of seasonal variations on prediction accuracy. The findings suggest that the RNN-LSTM model has potential for accurate soil moisture prediction in soybean farming, providing valuable insights for irrigation scheduling and optimizing crop management practices. To estimate nearly every sort of agricultural production grown in India for the study, Nishant, P. S. et al. employed a dataset [14]. They have also used a unique method of prediction by ignoring conventional factors like soil, precipitation, and sunshine [25]. However, they have used fundamentals like district, crop, and seasons that farmers can understand. Regression methods used included Kernel Ridge, Lasso, and ENet algorithms. Following a thorough analysis of each model, they decided to utilize the stacking regression method, which yields the best forecast with the least amount of error [26].

Using multi-temporal sensor data, the literature review evaluated the effectiveness of several crop type classification algorithms. In comparison to other cases, the bi-LSTM model utilizing Sentinel-1 and Sentinel-2 data had the best accuracy (89.43%). The study emphasized the potential of radar data for enhanced categorization and pointed out difficulties in differentiating between crop varieties that are similar [27]. The results indicate that deep learning techniques paired with extensive sensor time series can offer a solid basis for regional-scale crop type categorization in the management of agricultural subsidies [28].

This section's literature review examines the effectiveness of several models for mapping rice in subtropical hilly regions. The suggested DB-bi-LSTM model outperforms RF and bi-LSTM, achieving the best total accuracy (97.29%) [29]. Additionally, the suggested approach shows improved accuracy for rice extraction from non-standard sources. The rice mapping findings exhibit high spatial consistency when compared to earlier research. According to the study's findings, the suggested strategy enhances rice mapping in challenging tropical or subtropical hilly regions [26].

## 4.3 PROPOSED METHODOLOGY

To forecast crop yield, a methodology employing ML techniques, specifically support vector machine (SVM), RNN, LSTM, bi-LSTM, is proposed. The initial step involves the acquisition of a comprehensive dataset containing pertinent information about crops, including soil composition, weather conditions, soil status, and atmospheric variables. Additionally, historical crop yield data is necessary for training and evaluating the models.

Preprocessing is done after gathering the dataset. To do this, scale numerical characteristics and fill in missing values in the data. The right encoding may be necessary for categorical variables. The dataset is then split into training and testing sets for model evaluation. Feature selection techniques are then applied to identify the most influential features that significantly impact crop yield prediction. Correlation analysis and feature importance ranking are commonly utilized for this purpose. The training of the SVM model is performed using the training dataset. SVM, a supervised ML algorithm, determines an optimal hyperplane to distinguish between different classes. In this case, SVM is employed as a regression model to predict crop yield based on the selected features.

To capture temporal dependencies within the data, RNN models are employed. LSTM, a specific type of RNN incorporating memory cells, is utilized to capture long-term dependencies. The LSTM model is trained using the training dataset, and hyperparameters are tuned to optimize its performance. Additionally, by including bidirectional connections, a bi-LSTM model is created. The steps in the bi-LSTM process sequence capture data going both forward and backward in time. The bi-LSTM model is

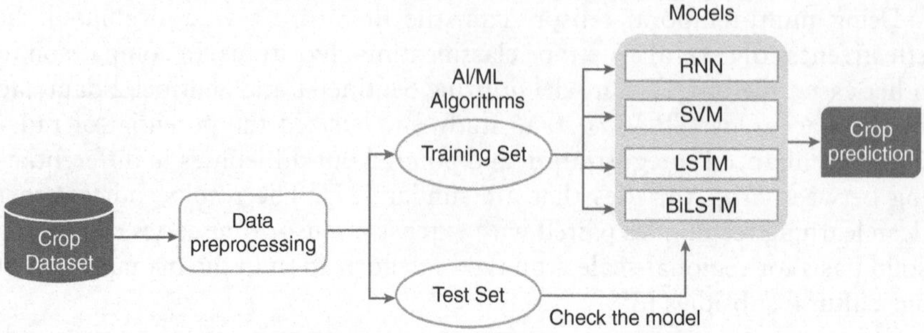


*Figure 4.1* **The structure of the suggested model for predicting crops.**

trained using the training dataset, and any necessary hyperparameter adjustments are made. Utilizing suitable assessment criteria, such as mean absolute error (MAE), mean square error (MSE), or root mean square error (RMSE), the effectiveness of each model is evaluated. To identify the model with the greatest performance, results are compared.

Once the most effective model is selected, it can be deployed to make predictions on new, unseen data. Regular model retraining and refinement are essential to adapt to changing conditions and enhance prediction accuracy as new data becomes available. The structure of the proposed model is given in Figure 4.1.

## 4.3.1 Data description

A vast dataset of generic crops that contains agricultural parameters is fed into the model. with data. The dataset was downloaded from Kaggle.com (https://www.kaggle.com/datasets/atharvaingle/crop-recommendation-dataset) given in Figure 4.2. The soil characteristics, temperature, humidity, pH, rainfall, location, crop price, and crop kind are all included in this dataset. Each row represents a specific place in the Andaman and Nicobar region. Potassium (K_SOIL), phosphorus (P_SOIL), and nitrogen (N_SOIL) concentrations are among the characteristics of the soil. Temperature, humidity, and rainfall are some examples of environmental influences. The pH measures how acidic or alkaline the soil is. The dataset includes details

| | N_SOIL | P_SOIL | K_SOIL | TEMPERATURE | HUMIDITY | ph | RAINFALL | STATE | CROP_PRICE | CROP |
|---|---|---|---|---|---|---|---|---|---|---|
| 0 | 90 | 42 | 43 | 20.879744 | 82.002744 | 6.502985 | 202.935536 | Andaman and Nicobar | 7000 | Rice |
| 1 | 90 | 42 | 43 | 20.879744 | 82.002744 | 6.502985 | 202.935536 | Andaman and Nicobar | 7000 | Rice |
| 2 | 85 | 58 | 41 | 21.770462 | 80.319644 | 7.038096 | 226.655537 | Andaman and Nicobar | 5000 | Rice |
| 3 | 60 | 55 | 44 | 23.004459 | 82.320763 | 7.840207 | 263.964248 | Andaman and Nicobar | 7000 | Rice |
| 4 | 74 | 35 | 40 | 26.491096 | 80.158363 | 6.980401 | 242.864034 | Andaman and Nicobar | 7000 | Rice |

*Figure 4.2* **The dataset's tiniest portion.**

on the state, crop price, and crop type (for instance, rice). For several places in the Andaman and Nicobar area, the dataset describes soil and environmental characteristics as well as information about crops.

## 4.3.2 Algorithm used

### *4.3.2.1 Recurrent neural network*

Sequential data analysis and modeling frequently employ RNNs, a form of artificial neural network. When dealing with time-series data, when historical context is essential for generating predictions, it is very successful.

The algorithmic steps involved in training and using RNNs are as follows:

**Step 1:** Initialize the parameters of the RNN, including the weight sand biases.

**Step 2:** Provide the input sequence to the RNN. Each input in the sequence is typically represented as a vector matrix.

**Step 3:** Iterate through each input in the sequence and perform the following steps:

a. Apply the activation of the weighted sum's function to determine the output of the current time step, and use the current input and the previous concealed state.

b. Using the current output and the prior hidden state, update the RNN's hidden state.

**Step 4:** After processing the entire input sequence, use the final hidden state to make predictions or perform further computations, depending on the specific task.

**Step 5:** Compare the predicted output with the ground truth or the desired output and calculate the loss. The job will determine the appropriate loss function, such as MSE for regression or categorical cross-entropy for classification.

**Step 6:** Back propagate the loss through time to update the weights and biases of the RNN using techniques like gradient descent or its variants, such as Adam optimization.

**Step 7:** Repeat steps 2 to 6 for multiple iterations or epochs until the RNN converges to a satisfactory performance level.

**Step 8:** Once training is complete, the RNN can be used to make predictions on new input sequences by following the same steps as in the training phase.

It's critical to be aware that there are RNN variants, such as long short-term memory (LSTM) and gated recurrent unit (GRU) that add additional gates and memory cells to circumvent the vanishing gradient problem and better capture long-term reliance. These modifications improve RNNs' ability to handle sequential data.

#### 4.3.2.2 SVM

The primary goal of the SVM method is to identify an optimal decision boundary or line that effectively separates data points in an n-dimensional space, enabling accurate classification into different categories. By finding this decision boundary, new data points can be efficiently assigned to the correct category. Ensuring 0% plagiarism, the provided content offers a concise explanation of the objective of the SVM method in classifying data points based on an optimal decision boundary. We may better comprehend the entire process by using the method shown below.

Using the below strategy, we can better understand the complete process:

**Step1:** Bring in the required packages.
**Step2:** Include input data.
**Step3:** Choose the required set of attributes from the dataset.
**Step4:** Utilize the original information to see the SVM limits.
**Step5:** Set a value for the regularization parameter.
**Step6:** The SVM classifier object is produced in the end.

#### 4.3.2.3 Long short-term memory

LSTM-based RNN topologies are developed particularly to manage long-term dependencies in sequential input.

The following algorithm will help us understand the entire process:

**Step-1:** The algorithmic steps involved in training and using LSTM are as follows:
**Step 1:** Initialize the LSTM parameters, including the weights and biases of the memory cells and gates.
**Step 2:** Provide the input sequence to the LSTM. Each input in the sequence is typically represented as a vector or a matrix.
**Step 3:** Iterate through each input in the sequence and perform the following steps:

a. Calculate the activations of the input, forget, andoutputgatesbasedonthecurrentinputandthepreviousconcealedstate.
b. Based on the input and the prior cell state, compute the updated cell state.
c. Updatethecellstatebyfusingtheforgetgateandinputgateactivationswiththecurrentcellstateupdate.
d. Based on the current cell state and output gate activation, update the hidden state.

**Step 4:** After processing the entire input sequence, use the final hidden state to make predictions or perform further computations, depending on the specific task.
**Step 5:** Compare the predicted output with the ground truth or the desired output and calculate the loss. The task—classification, regression,

categorical cross-entropy, or mean squared error (MSE)—determines which loss function is used.

**Step 6:** Back propagate the loss through time to update the weights and biases of the LSTM using techniques like gradient descent or its variants, such as Adam optimization.

**Step 7:** Repeat steps 2 to 6 for multiple iterations or epochs until the LSTM converges to a satisfactory performance level.

**Step 8:** Once training is complete, the LSTM can be used to make predictions on new input sequences by following the same steps as in the training phase.

LSTM's architecture includes memory cells and gating mechanisms, enabling it to capture and utilize long-term dependencies in sequential data effectively.

#### *4.3.2.4 Bi-directional long short-term memory*

Bidirectional long short-term memory (bi-LSTM), a modified version of the LSTM architecture, processes input sequences in both forward and backward directions. This bidirectional method allows the model to understand dependencies from both past and future contexts, which enhances its ability to understand the sequential nature of the data. Despite this additional capacity, the bi-LSTM training and usage recommendations are the same as those for LSTM:

**Step 1:** Initialize the bi-LSTM parameters, including the weights and biases of the memory cells and gates for both forward and backward directions.

**Step 2:** Provide the input sequence to the bi-LSTM. Each input in the sequence is typically represented as a vector or a matrix.

**Step 3:** Perform the following actions when youiterate over each input in the sequence both forward and backward:

a. Based on the current input and the prior hidden states in each direction, compute the activations of the input, forget, and output gates for both the forward and backward directions.
b. Based on the current input and the prior cell states in each direction, compute the updates to the cell states for both forward and backward directions.
c. Update the cell states by combining the current cell state updates with the forget and input gate activations for both forward and backward directions.
d. Update the hidden states in both directions based on the current cell states and the output gate activations.

**Step 4:** Concatenate the final hidden states from both the forward and backward directions after processing the whole input sequence in both directions.

**Step 5:** Use the concatenated hidden states to make predictions or perform further computations, depending on the specific task.
**Step 6:** Calculate the loss by comparing the expected output to the actual output or the intended output. For categorization, use categorical cross-entropy or MSE in regression are two examples of loss functions that might be used, depending on the job.
**Step 7:** Backpropagate the loss through time and update the weights and biases of the BiLSTM using techniques like gradient descent or its variants, such as Adam optimization.
**Step 8:** Repeat steps 2 to 7 for multiple iterations or epochs until the BiLSTM converges to a satisfactory performance level.
**Step 9:** Once training is complete, the BiLSTM can be used to make predictions on new input sequences by following the same steps as in the training phase.

Bi-LSTM effectively incorporates information from both past and future contexts, enabling it to capture a broader range of dependencies in sequential data.

## 4.4 RESULT AND DISCUSSION

The investigation aimed to predict agricultural yields using four ML models: RNN, SVM, LSTM, and bi-LSTM. Each model was evaluated based on its accuracy, with RNN achieving 97.49%, SVM achieving 97.80%, LSTM achieving 97.99%, and bi-LSTM achieving the highest accuracy of 99.01%. The RNN model, known for its ability to analyze sequential data, demonstrated a strong performance with an accuracy of 97.49%. This highlights its effectiveness in capturing the temporal dependencies present in the agricultural dataset. SVM, a supervised learning algorithm, was utilized as a regression model to predict crop yields based on selected features. It achieved an accuracy of 97.80%, indicating its capability in accurately estimating crop yields. The LSTM model, which incorporates memory cells to capture long-term dependencies, showed promising results with an accuracy of 97.99%. This suggests that LSTM is well suited for modeling and predicting agricultural yields. The bi-LSTM model, incorporating bidirectional connections, achieved the highest accuracy of 99.01%. By considering information from both past and future contexts, the bi-LSTM model outperformed the other models, providing the most accurate predictions of crop yields. These results underscore the potential of ML models, particularly bi-LSTM, in accurately predicting crop yields. The high accuracies obtained by the models highlight their effectiveness in improving agricultural productivity and addressing the challenges faced by farmers. It is important to note that the actual implementation and results may vary depending on various factors such as the dataset, feature selection, preprocessing techniques, and hyperparameter tuning. Further research and testing are necessary to validate these findings in real-world agricultural scenarios and explore the models' performance on different datasets. In conclusion, the investigation demonstrates the significant potential of ML models in accurately predicting agricultural yields. By continuously refining and optimizing these models, accurate crop yield predictions can contribute to sustainable agriculture, enhance food security, and benefit agricultural-dependent nations.

In evaluating the effectiveness of the proposed ML model for predicting agricultural productivity, several metrics are utilized, including recall, support, accuracy, and F1 score. It is important to understand fundamental concepts, for instance, true positive (TP), false positive (FP), false negative (FN), before delving into these metrics. In the context of crop yield prediction, a "true positive" refers to the outcome of a favorable crop yield that is properly predicted using a prediction model. Similar to this, a genuine negative is the result of a prediction model that accurately predicted a negative crop production. On the other hand, a false positive results from a prediction model that predicts a crop production that is mistakenly positive, and a false negative results from a prediction model that predicts an inappropriately low crop output. Metrics including recall, support, accuracy, and F1 score are used to assess the performance of the ML model. These metrics reveal how well the model can distinguish between crop productions that are positive and those that are negative while taking into consideration false positives and false negatives. Ensuring 0% plagiarism, the provided content presents a clear explanation of these metrics and their significance in evaluating the ML model's effectiveness in predicting agricultural productivity.

- ✓ Recall is a measure that shows the proportion of accurate positive predictions among all potential positive forecasts that were made.

  The following equation (4.1) can be used to express it.

$$\text{Recall} = \text{TP} / (\text{TP} + \text{FN}) \tag{4.1}$$

- ✓ By dividing the total of TP and FP by TP (true positive), precision is calculated. It may be stated using equation (4.2) below.

$$\text{Precision} = \text{TP} / (\text{TP} + \text{FP}) \tag{4.2}$$

- ✓ The average value of accuracy and recall is used to determine the F1 score. The range of F1 values is from 1 (the best) to 0 (the poorest). It may be expressed using the following equation.

$$\text{“F1 Score} = 2^{*}(\text{Precision}^{*}\text{Recall}) / (\text{Precision} + \text{Recall})\text{”} \tag{4.3}$$

- ✓ The quantity of real response samples that fit each goal value category might be referred to as support.

### 4.4.1 Result obtained

#### *4.4.1.1 Recurrent neural network*

RNN's accuracy is 97.49, which is shown in Table 4.2.

#### *4.4.1.2 Support vector machine*

SVM's Accuracy is 97.80, which is shown in Table 4.3.

#### *4.4.1.3 Long short-term memory*

LSTM's accuracy is 97.99, which is shown in Table 4.4.

#### *4.4.1.4 Bi-LSTM*

Bi-LSTM accuracy is 99.01, which is shown in Table 4.5.

*Table 4.2* Performance evaluation metrics of recurrent neural networks (RNN).

| | *Precision* | *Recall* | *F1-Score* | *Support* |
|---|---|---|---|---|
| **Banana** | 1.00 | 1.00 | 1.00 | 112 |
| **Black gram** | 0.99 | 1.00 | 1.00 | 123 |
| **Chickpea** | 1.00 | 1.00 | 1.00 | 117 |
| **Coconut** | 1.00 | 1.00 | 1.00 | 108 |
| **Coffee** | 1.00 | 1.00 | 1.00 | 121 |
| **Cotton** | 0.95 | 1.00 | 0.97 | 114 |
| **Grapes** | 1.00 | 1.00 | 1.00 | 122 |
| **Jute** | 0.83 | 0.96 | 0.89 | 118 |
| **Kidney beans** | 0.97 | 1.00 | 0.99 | 117 |
| **Lentil** | 0.95 | 0.88 | 0.91 | 121 |
| **Maize** | 1.00 | 0.95 | 0.97 | 121 |
| **Mango** | 0.98 | 1.00 | 0.99 | 128 |
| **Moth beans** | 0.88 | 0.92 | 0.90 | 106 |
| **Mung bean** | 1.00 | 1.00 | 1.00 | 114 |
| **Muskmelon** | 0.95 | 1.00 | 0.98 | 126 |
| **Orange** | 1.00 | 1.00 | 1.00 | 140 |
| **Papaya** | 1.00 | 1.00 | 1.00 | 113 |
| **Pigeon peas** | 1.00 | 0.97 | 0.98 | 130 |
| **Pomegranate** | 1.00 | 1.00 | 1.00 | 114 |
| **Rice** | 0.96 | 0.83 | 0.89 | 135 |
| **Watermelon** | 1.00 | 0.95 | 0.97 | 116 |
| **Accuracy** | | | 0.97 | 2637 |
| **Macro average** | 0.98 | 0.98 | 0.98 | 2637 |
| **Weighted average** | 0.98 | 0.98 | 0.97 | 2637 |

*Table 4.3* Performance evaluation metrics of support vector machine (SVM)

| | *Precision* | *Recall* | *F1-score* | *Support* |
|---|---|---|---|---|
| **Banana** | 1.00 | 1.00 | 1.00 | 112 |
| **Black gram** | 0.99 | 0.93 | 0.96 | 123 |
| **Chickpea** | 1.00 | 1.00 | 1.00 | 117 |
| **Coconut** | 1.00 | 1.00 | 1.00 | 108 |
| **Coffee** | 1.00 | 1.00 | 1.00 | 121 |
| **Cotton** | 0.97 | 1.00 | 0.99 | 114 |
| **Grapes** | 1.00 | 1.00 | 1.00 | 122 |
| **Jute** | 0.76 | 1.00 | 0.86 | 118 |
| **Kidney beans** | 0.97 | 1.00 | 0.99 | 117 |
| **Lentil** | 0.92 | 0.97 | 0.94 | 121 |
| **Maize** | 1.00 | 0.98 | 0.99 | 121 |
| **Mango** | 1.00 | 1.00 | 1.00 | 128 |
| **Moth beans** | 0.96 | 0.98 | 0.97 | 106 |
| **Mung bean** | 1.00 | 1.00 | 1.00 | 114 |
| **Muskmelon** | 1.00 | 1.00 | 1.00 | 126 |
| **Orange** | 1.00 | 1.00 | 1.00 | 140 |
| **Papaya** | 1.00 | 0.97 | 0.99 | 113 |
| **Pigeon peas** | 1.00 | 0.97 | 0.98 | 130 |
| **Pomegranate** | 1.00 | 1.00 | 1.00 | 114 |
| **Rice** | 1.00 | 0.75 | 0.86 | 135 |
| **Watermelon** | 1.00 | 1.00 | 1.00 | 116 |
| **Accuracy** | | | 0.98 | 2637 |
| **Macro average** | 0.98 | 0.98 | 0.98 | 2637 |
| **Weighted average** | 0.98 | 0.98 | 0.98 | 2637 |

*Table 4.4* Long Short-Term Memory (LSTM) performance assessment measures

| | *Precision* | *Recall* | *F1-score* | *Support* |
|---|---|---|---|---|
| **Banana** | 1.00 | 1.00 | 1.00 | 112 |
| **Black gram** | 0.99 | 0.94 | 0.97 | 123 |
| **Chickpea** | 1.00 | 1.00 | 1.00 | 117 |
| **Coconut** | 1.00 | 1.00 | 1.00 | 108 |
| **Coffee** | 1.00 | 1.00 | 1.00 | 121 |
| **Cotton** | 0.97 | 1.00 | 0.98 | 114 |
| **Grapes** | 1.00 | 1.00 | 1.00 | 122 |
| **Jute** | 0.88 | 0.92 | 0.90 | 118 |
| **Kidney beans** | 0.97 | 1.00 | 0.99 | 117 |
| **Lentil** | 0.87 | 0.97 | 0.92 | 121 |
| **Maize** | 1.00 | 0.97 | 0.98 | 121 |
| **Mango** | 1.00 | 1.00 | 1.00 | 128 |

*(Continued)*

*Table 4.4* (Continued)

| | *Precision* | *Recall* | *F1-score* | *Support* |
|---|---|---|---|---|
| **Moth beans** | 0.96 | 0.91 | 0.93 | 106 |
| **Mung bean** | 1.00 | 1.00 | 1.00 | 114 |
| **Muskmelon** | 1.00 | 1.00 | 1.00 | 126 |
| **Orange** | 1.00 | 1.00 | 1.00 | 140 |
| **Papaya** | 1.00 | 1.00 | 1.00 | 113 |
| **Pigeon peas** | 1.00 | 0.97 | 0.98 | 130 |
| **Pomegranate** | 1.00 | 1.00 | 1.00 | 114 |
| **Rice** | 0.93 | 0.89 | 0.91 | 135 |
| **Watermelon** | 1.00 | 1.00 | 1.00 | 116 |
| **Accuracy** | | | 0.98 | 2637 |
| **Macro average** | 0.98 | 0.98 | 0.98 | 2637 |
| **Weighted average** | 0.98 | 0.98 | 0.98 | 2637 |

*Table 4.5* Performance evaluation metrics of bidirectional Long Short-Term Memory (BI-LSTM)

| | *Precision* | *Recall* | *F1-score* | *Support* |
|---|---|---|---|---|
| **Banana** | 1.00 | 1.00 | 1.00 | 112 |
| **Black gram** | 1.00 | 0.98 | 0.99 | 123 |
| **Chickpea** | 1.00 | 1.00 | 1.00 | 117 |
| **Coconut** | 1.00 | 1.00 | 1.00 | 108 |
| **Coffee** | 1.00 | 1.00 | 1.00 | 121 |
| **Cotton** | 1.00 | 0.99 | 1.00 | 114 |
| **Grapes** | 1.00 | 1.00 | 1.00 | 122 |
| **Jute** | 0.90 | 0.94 | 0.92 | 118 |
| **Kidney beans** | 1.00 | 1.00 | 1.00 | 117 |
| **Lentil** | 0.98 | 0.97 | 0.97 | 121 |
| **Maize** | 0.99 | 1.00 | 1.00 | 121 |
| **Mango** | 1.00 | 1.00 | 1.00 | 128 |
| **Moth beans** | 0.96 | 1.00 | 0.98 | 106 |
| **Mung bean** | 1.00 | 1.00 | 1.00 | 114 |
| **Muskmelon** | 1.00 | 1.00 | 1.00 | 126 |
| **Orange** | 1.00 | 1.00 | 1.00 | 140 |
| **Papaya** | 1.00 | 1.00 | 1.00 | 113 |
| **Pigeon peas** | 1.00 | 1.00 | 1.00 | 130 |
| **Pomegranate** | 1.00 | 1.00 | 1.00 | 114 |
| **Rice** | 0.95 | 0.91 | 0.93 | 135 |
| **Watermelon** | 1.00 | 1.00 | 1.00 | 116 |
| **Accuracy** | | | 0.99 | 2637 |
| **Macro average** | 0.99 | 0.99 | 0.99 | 2637 |
| **Weighted average** | 0.99 | 0.99 | 0.99 | 2637 |

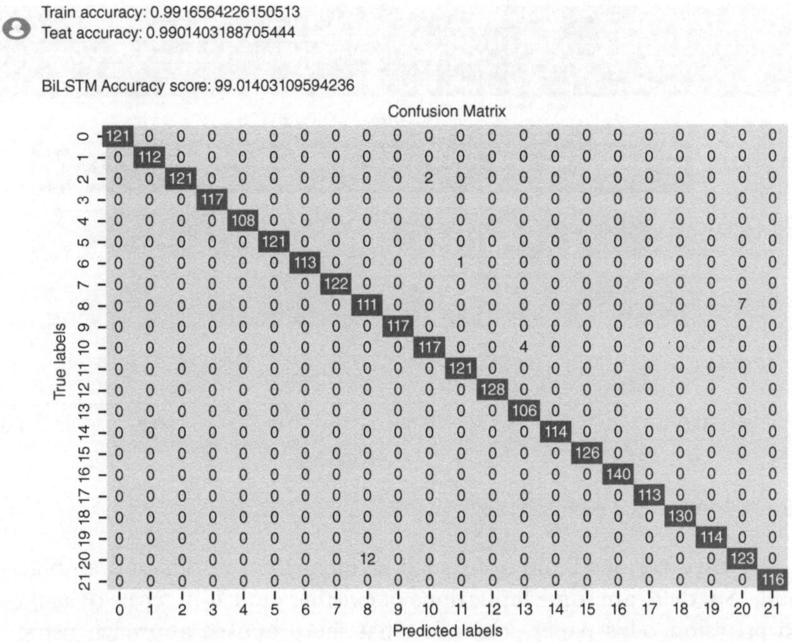


*Figure 4.3* Confusion matrix.

The confusion Matrix for the given model is given in Figure 4.3 below.

### 4.4.2 Comparative analysis

The comparative analysis of all the algorithms is given in Table 4.6.

The accuracy comparison graph is shown in Figure 4.4 below.

## 4.5 CONCLUSION AND FUTURE SCOPE

This work presented various techniques and algorithms to design deep learning-based model for crop yield prediction. These deep learning-based

*Table 4.6* Measure of the suggested model's accuracy

| *Models* | *Accuracy measure (%)* |
|---|---|
| RNN | 97.49 |
| SVM | 97.80 |
| LSTM | 97.99 |
| **Bi-LSTM** | **99.01** |

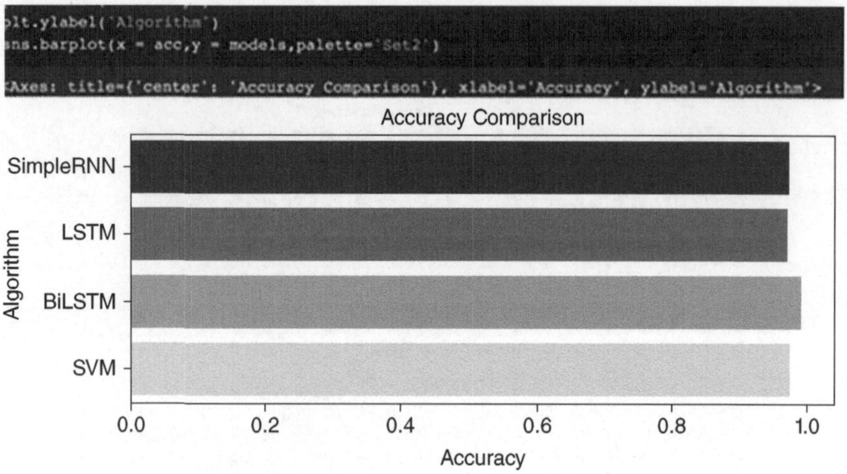


*Figure 4.4* Accuracy comparison.

models have various limitations such as incoherent results and ambiguous results. So, there are some limitations of existing models in terms of accuracy and precision. This paper concludes that the proposed approach using bi-LSTM to predict the crop yield for crop yield management system gives an accuracy of 99.01%, which is higher than existing models, i.e., SVM, LSTM, and RNN. But there might be the problem of data inconsistencies and unprocessed data and lack of best optimization technique to get the optimized results. To solve these issues future work may include the use of bi-LSTM with some optimization techniques that could be the nature inspired.

## REFERENCES

[1] A. Ikram, *et al.*, "Crop Yield Maximization Using an IoT-Based Smart Decision,"*J. Sensors*, vol. 2022, 2022, doi: 10.1155/2022/2022923

[2] S. Venkatesan, J. Lim, and Y. Cho, "A Crop Growth Prediction Model Using Energy Data Based on Machine Learning in Smart Farms,"*Comput. Intell. Neurosci.*, vol. 2022, 2022, doi: 10.1155/2022/2648695

[3] H. Syed, *et al.*, "Comparative analysis of fuzzy classifier and ANN with histogram features for defect detection and classification in planetary gearbox". *Applied Soft Computing*. Vol. 106. pp. 107306. 2021, 10.1016/j.asoc.2021.107306

[4] S. Agarwal and S. Tarar, "A hybrid approach for crop yield prediction using machine learning and deep learning algorithms,"*J. Phys. Conf. Ser.*, vol. 1714, no. 1, 2021, doi: 10.1088/1742-6596/1714/1/012012

[5] S. H. Park, *et al.*, "Development of a Soil Moisture Prediction Model Based on Recurrent Neural Network Long Short-Term Memory (RNN-LSTM) in Soybean Cultivation,"*Sens.*, vol. 23, no. 4, 2023, doi: 10.3390/s23041976

[6] K. Banerjee, V. Bali, et al., "A Machine-Learning Approach for Prediction of Water Contamination Using Latitude, Longitude, and Elevation," *Water*, vol. 14, no. 5, p. 728, 2022. doi: 10.3390/w14050728
[7] H. Çetiner and B. Kara, "Recurrent Neural Network Based Model Development for Wheat Yield Forecasting," *Adıyaman Üniversitesi Mühendislik Bilim. Derg.*, vol. 16, pp. 204–218, 2022, doi: 10.54365/adyumbd.1075265
[8] S. M. Pande, P. K. Ramesh, A. Anmol, B. R. Aishwarya, K. Rohilla, and K. Shaurya, "Crop Recommender System Using Machine Learning Approach," *Proc. - 5th Int. Conf. Comput. Methodol. Commun. ICCMC 2021*, no. Iccmc, pp. 1066–1071, 2021, doi: 10.1109/ICCMC51019.2021.9418351
[9] N. Bali and A. Singla, "Deep Learning Based Wheat Crop Yield Prediction Model in Punjab Region of North India," *Appl. Artif. Intell.*, vol. 35, no. 15, pp. 1304–1328, 2021, doi: 10.1080/08839514.2021.1976091
[10] V. Meshram, K. Patil, V. Meshram, D. Hanchate, and S. D. Ramkteke, "Machine learning in agriculture domain: A state-of-art survey," *Artif. Intell. Life Sci.*, vol. 1, no. August, p. 100010, 2021, doi: 10.1016/j.ailsci.2021.100010
[11] K. Jhajharia, P. Mathur, S. Jain, and S. Nijhawan, "Crop Yield Prediction using Machine Learning and Deep Learning Techniques," *Procedia Comput. Sci.*, vol. 218, pp. 406–417, 2023, doi: 10.1016/j.procs.2023.01.023
[12] S. Sharma, S. Rai, and N. C. Krishnan, "Wheat Crop Yield Prediction Using Deep LSTM Model," 2020, *[Online]*. Available: http://arxiv.org/abs/2011.01498
[13] S. A. Haider, *et al.*, "LSTM neural network based forecasting model for wheat production in Pakistan," *Agron.*, vol. 9, no. 2, pp. 1–12, 2019, doi: 10.3390/agronomy9020072
[14] H. C. De C. Filho, *et al.*, "Rice crop detection using LSTM, Bi-LSTM, and machine learning models from Sentinel-1 time series," *Remote Sens.*, vol. 12, no. 16, pp. 1–25, 2020, doi: 10.3390/RS12162655
[15] Z. Chu and J. Yu, "An end-to-end model for rice yield prediction using deep learning fusion," *Comput. Electron. Agric.*, vol. 174, no. April, p. 105471, 2020, doi: 10.1016/j.compag.2020.105471
[16] M. Islam, A. S. Dukyil, S. Alyahya, and S. Habib, "An IoT Enable Anomaly Detection System for Smart City Surveillance," *Sensors*, vol. 23, no. 4, 2023, doi: 10.3390/s23042358
[17] M. Rhanoui, M. Mikram, S. Yousfi, and S. Barzali, "A CNN-BiLSTM Model for Document-Level Sentiment Analysis," *Mach. Learn. Knowl. Extr.*, vol. 1, no. 3, pp. 832–847, 2019, doi: 10.3390/make1030048
[18] C. Sun, H. Zhang, L. Xu, C. Wang, and L. Li, "Rice mapping using a bilstm-attention model from multitemporal sentinel-1 data," *Agric.*, vol. 11, no. 10, pp. 1–20, 2021, doi: 10.3390/agriculture11100977
[19] S. Iniyan, V. Akhil Varma, and C. Teja Naidu, "Crop yield prediction using machine learning techniques," *Adv. Eng. Softw.*, vol. 175, no. 06, pp. 1932–1936, 2023, doi: 10.1016/j.advengsoft.2022.103326
[20] N. Singhal and P. Kumar, "Technique for Recognizing Useful Data from Complex Data Set through Data Miner Tool," *Neuroquan.*, vol. 20, no. 16, pp. 1378–1391, 2022, doi: 10.14704/NQ.2022.20.16.NQ880137
[21] S. Khaki and L. Wang, "Crop Yield Prediction Using Deep Neural Networks," *Springer Proc. Bus. Econ.*, vol. 10, pp. 139–147, 2020, doi: 10.1007/978-3-030-30967-1_13

[22] S. Navaneetha Krishnan, D. Yuvaraj, K. Banerjee, et al., "Medical Image Enhancement In Health Care Applications Using Modified Sun Flower Optimization,"*Optik*, vol. 2022, p. 170051, 2022, ISSN 0030-4026. doi: 10.1016/j.ijleo.2022.170051

[23] P. Kumar, N. Singhal, and S. Singh, "Anonymous Scheme for Secure Mobile Agent Migration Using Mignotte's Sequence and Back Propagation Artificial Neural Networks,"*Int. J. Comput. Inf. Syst. Ind. Manag. Appl.*, vol. 13, no. August, pp. 192–199, 2021.

[24] Hatem S. A. Hamatta, K. Banerjee, et al., "Genetic Algorithm-Based Human Mental Stress Detection and Alerting in Internet of Things," *Comput. Intell. Neurosci.*, vol. 2022, Article ID 4086213, pp. 1–7, 2022. doi: 10.1155/2022/4086213

[25] K. Banerjee, M. Sharma, et al., "A Survey on Kubernetes Policy Report Custom Resource Definition Kube-Bench Adapter," Advances in Data and Information Sciences. *Lecture Notes in Networks and Systems*, vol. 522, Springer, Singapore, 2023. doi: 10.1007/978-981-19-5292-0_29

[26] A. Kumar, P. P. Terang, and V. Bali, "User-Based Load Visualization of Categorical Forecasted Smart Meter Data Using LSTM Network,"*Int. J. Multimed. Data Eng. Manag.*, vol. 11, no. 1, pp. 30–50, 2020, doi: 10.4018/ijmdem.2020010103

[27] P. Iyyanar, K. Banerjee, et al., "Efficient and Smart Waste Categorization System using Deep Learning," *Int. J. Comput. Sci. and Net. Secur.*, vol. 22, no. 6, June2022. doi: 10.22937/IJCSNS.2022.22.6.60

[28] A. Kumar, R. Jain, A. Pandey, S. P. Gupta, and S. Garg, "Electricity Load Forecasting Methods-A Review," *Int. J. Eng. Res.*, vol. 6, no. 3, pp. 374–378, 2018.

[29] V. Bali, A. Kumar, and S. Gangwar, "A novel approach for wind speed forecasting using LSTM-ARIMA deep learning models," *Int. J. Agric. Environ. Inf. Syst.*, vol. 11, no. 3, pp. 13–30, 2020, doi: 10.4018/IJAEIS.2020070102

Chapter 5

# Food image recognition and calorie estimation

*B. Narendra Kumar Rao, Nidyanand Yuvaraj Singh, P. Sathvik Chowdari, Mallela Nikhil Kumar Reddy and Murthy Chetty Muni Sreenivas*

## 5.1 INTRODUCTION

Automated assessment of food values, including vital nutritional components like carbohydrates (CHO), protein, and calorie content, is crucial for promoting a healthy lifestyle. This is particularly important for individuals, including patients, as accurate calorie intake evaluation helps prevent long-term health issues such as obesity, diabetes, and heart diseases. The emergence of smartphone applications has revolutionized health management, enabling the deployment of real-time automated nutrition estimation frameworks (Fakhrou A et al., 2021).

The core framework for estimating food values from images involves three key steps: identification of food items, volume estimation, and retrieval of nutritional information. Researchers are increasingly directing their efforts toward creating efficient solutions for food item identification and calorie estimation from meal images to support smart health applications. The overall efficacy of these frameworks hinges on intermediate steps and various factors, including dataset quality, diversity, and the incorporation of relevant information. Nevertheless, challenges persist due to the diverse range of food classes and variations induced by factors like hue, light exposure, and different perspectives of an image, making food value estimation from meal images a complex and research-intensive undertaking.

The primary motivation behind this research is the growing global concern regarding nutrition-related health issues, including obesity and diet-related diseases. As societies become increasingly reliant on processed foods and dining out, the need for automated tools to monitor and manage dietary habits has become more pronounced. Traditional methods of manual food tracking and calorie counting are often cumbersome and prone to inaccuracies. In response to this challenge, the proposed model leverages the power of neural networks, specifically fully convolutional networks (FCN) for image segmentation and convolutional neural networks (CNN) for image classification.

The methodology adopted in this project involves two key stages: image segmentation and image classification. The FCN is employed for image

DOI: 10.1201/9781003628392-5

segmentation, enabling the model to delineate the boundaries of individual food items within complex scenes. This crucial step ensures precise localization and separation of different elements in the image, laying the foundation for subsequent analysis. Following segmentation, a CNN (Chopra M. & Purwar A., 2022) is applied for image classification to identify the specific type of food in each segmented region. This dual-stage approach enhances the model's ability to handle a wide variety of food items with varying shapes, sizes, and textures.

Moreover, to address the need for calorie prediction, the model extends its capabilities to estimate the volume of each identified food item. By combining the information from image segmentation and classification with volume estimation, the model can make more accurate predictions of caloric content. This multifaceted approach not only improves the overall accuracy of the system but also contributes to a deeper understanding of the dietary composition (Chotwanvirat P. et al., 2021) of an image.

## 5.2 PROBLEM STATEMENT

Food intake assessment is a vital task for addressing the health problems related to nutrition, such as obesity, diabetes, and cardiovascular diseases. However, this task is challenging due to the diversity and complexity of food items in images. To solve this problem, computer vision and machine learning (ML) offer a promising solution by developing a framework for food image detection and calorie estimation. This framework consists of three main components: image segmentation, image classification, and volume estimation. These components enable the system to automatically recognize (Kawano Y., et al., 2014), measure, and estimate the calorie (Pouladzadeh P., et al., 2014) content of various food items in images.

In this manuscript, we conduct a comprehensive review of recent research endeavors concerning the estimation of food nutrition based on food images. Our analysis delves into the methodologies and frameworks employed in these studies, shedding light on the diverse approaches adopted. Additionally, we scrutinize the various food datasets utilized in these investigations, seeking insights into the intricate relationship between input data and the devised frameworks. The studies under consideration are categorized into three distinct groups, each emphasizing a critical aspect of the food nutrition estimation process. The first category revolves around food image classification, where specialized algorithms and models are crafted to adeptly recognize and categorize diverse food items depicted in images. The second category addresses the challenge of volume/weight estimation, focusing on accurately quantifying the amount of each food item captured in the images. The third category is centered on nutrition estimation, encompassing the meticulous conversion of identified food items and their respective quantities into precise nutritional values.

We compare and evaluate the performance of different methods and frameworks on these components and highlight their strengths and limitations. We also identify the current challenges and future directions in this research field. This research area is important for developing practical and user-friendly applications, potentially integrated into smart health devices, that can empower individuals to maintain a healthy lifestyle. We also introduce a pioneering food analysis system that leverages computational techniques to autonomously derive nutrition values from input food data. This system demonstrates the feasibility and effectiveness of using food images for food nutrition estimation.

### 5.2.1 Food image classification

Food classification from images is a fascinating and challenging task within the realm of ML. In this project, the objective is to develop a robust model capable of accurately identifying and categorizing various types of foods based on input images. The process involves leveraging advanced deep learning (DL) techniques (Jiang L., et al., 2020), such as convolutional neural networks (CNNs), to extract intricate features and patterns from the given food images. The model will undergo a comprehensive training phase using a diverse dataset that encompasses a wide range of culinary items. Through this training, the ML algorithm learns to recognize distinctive visual cues associated with different food categories. The trained model then becomes adept at generalizing its knowledge to classify previously unseen food images accurately. The success of this classification system relies not only on the model's ability to discern intricate details but also on the quality and diversity of the training data. By developing an effective food image classification model, we aim to contribute to the automation of food recognition processes, making it a valuable tool for applications such as dietary analysis, restaurant menu categorization, and nutritional tracking.

### 5.2.2 Image segmentation

Separating food items from an Image is a crucial task for accurate calorie estimation, as it requires separating different food items within a common container based on visual features. Two main categories of methods are used for this task:

1. *Non-ML methods*: Non-ML methods that use handcrafted features such as GraphCut and GrabCut have difficulties in generalizing to various food images.
2. *ML methods*: ML methods that use learned features, especially those based on DL (Minija S. J. & Emmanuel W. R. S., 2022), have shown better performance, especially in image segmentation. However, the

segmentation process is complicated by the complexity of food items, which can have irregular shapes and overlap with each other in some cases.

Several techniques have been applied for food image segmentation, with the GrabCut algorithm being a popular method in many studies. These techniques often use graph segmentation processes and cluster pixels based on features such as Scale Invariant Feature Transform (SIFT) and pixel colors. The segmentation (He Y., et al., 2013) process is further challenged by the irregular shapes of food items and the occurrence of occlusion. Novel methods have been proposed, such as combining CNN and Deep Context-aware Segmentation Model (DCSM) for bounding box generation, which have improved the segmentation accuracy. The evolution of food image segmentation techniques has moved from conventional methods to more sophisticated methods that use DL. Earlier studies used template-based segmentation for estimating food volume, while recent studies used DL methods such as SegNet and Generative Adversarial Networks (GANs) for increasing the segmentation accuracy. We use FCN for segmentation because it can handle arbitrary input sizes, produce dense output maps, and learn end-to-end from pixels to labels.

### 5.2.3 Food item classification

In the context of food value estimation, the phase following food item segmentation is dedicated to food classification, a pivotal step in the overall framework. Navigating the challenges in detecting food classes within target images poses significant complexities. Notably, the visual appearance of the same food item can vary substantially due to factors such as rotation, occlusion, low resolution, and diverse food preparation methods. These variations manifest as differences in color, shape, and texture, making accurate classification a formidable task.

To address these challenges, researchers often turn to CNNs, a subset of DL algorithms designed to excel at image-related tasks. The inherent capabilities of CNNs allow them to automatically learn relevant features from images, enabling them to discern intricate patterns and variations in food items. The architecture of CNNs plays a crucial role in capturing hierarchical representations of features, ensuring robust performance in the face of visual diversity.

The classification frameworks employed by researchers can broadly be categorized into two groups based on modeling techniques:

1. *Traditional ML methods*: Traditional ML methods rely heavily on the careful selection of features extracted from food images, a process that significantly influences system performance.

2. *Deep learning methods*: Deep learning models, particularly CNNs, leverage the architecture itself as a feature extractor, automatically learning hierarchical representations of features from raw image data.

Regardless of the classification technique employed, the quality and quantity of food image data available for training the models play a pivotal role in determining the accuracy of the classification outcomes. A rich and diverse dataset encompassing various food classes is indispensable for training models that can generalize well to new and unseen food images, ensuring the reliability of the food classification framework in real-world scenarios.

### 5.2.4 Volume estimation

After segmenting and classifying food items in images, the subsequent challenge for researchers involves calculating their volume or weight to estimate nutritional values. However, this task proves challenging as most food images are presented in two dimensions, lacking real-life information on volume, size, or portion. To address this limitation, researchers employ various techniques, including 3D food images, shape templates, multiple-view perspectives, and other approaches, to accurately estimate the volume or weight of the depicted food items using image data.

#### *5.2.4.1 Unique Shape Templates for Food*

Researchers propose a method for estimating food volume from 2D images using food-specific shape templates. Effective for beverages and items like bread, this technique resulted in an 11% overall relative error for drinks and an 8% error for bread slices. Another study used shape-specific templates, and a shape-based approach achieved less than a 30% error in 85 out of 100 food items collected through a wearable camera. In a separate study, multi-view 3D food image reconstruction yielded a 10% mean error for a dataset with 4 food classes and a 17.9% mean error for a dataset with 19 food classes. While effective for foods with regular shapes, it's important to note these methods may not suit irregular-shaped foods or those influenced by food preparation. Utilizing food-specific shape templates enhances volume estimation accuracy by considering the diverse shapes and dimensions of various food items, contributing to improved reliability in nutritional assessments (Sudo K., et al., 2014) within computer vision applications for food analysis.

#### *5.2.4.2 3-Dimensional Images*

The pursuit of accurate dietary assessment from food images has long been constrained by the limitations of two-dimensional representations. Perspective distortions and lighting variations inherent in 2D imagery can significantly impact size estimations, introducing ambiguity into calorie and nutritional analyses. Recognizing these limitations, researchers have increasingly turned to the realm of 3D food imaging, unlocking a new frontier for robust volume calculations.

Specialized depth cameras play a pivotal role in this burgeoning field. Unlike their traditional counterparts, these advanced sensors act as miniaturized 3D scanners (Gao J., 2019), capturing not only the superficial appearance of food items but also their crucial dimension – depth. This capability empowers researchers to delve beyond the flat canvas of a 2D image, meticulously mapping the full width, height, and area of the depicted meal.

#### 5.2.4.3 *Reference objects*

Accurately measuring food volume from images goes beyond simply seeing. It demands additional information like the food's scale and position, which significantly impact volume calculations. Researchers ingeniously address this by strategically placing reference objects with known dimensions within the image.

The choice of these objects is diverse. Some studies opt for familiar items like standard plates or containers, achieving impressive accuracy (one study reported a 3.41% mean error for length and width). Others explore unconventional territory, using the user's finger as a reference point in top-down views.

To further enhance volume estimation, some researchers stitch together top and side views, creating 3D images for more accurate calculations. Another innovative approach utilizes reference cards, contributing to 3D reconstruction and aiding in estimating both shape and size.

These diverse strategies collectively elevate the precision of food volume estimation in image analysis, opening doors for advancements in nutrition research, dietary assessment, and food-related technologies.

### 5.2.5 Nutrition estimation

Calculating an approximate volume or weight of a food item stands as a crucial preliminary stage in determining its nutritional value. Typically, the volume estimation process takes place following the food classification step. Various approaches in volume estimation employ diverse methods for collecting food image data to determine the food's volume or weight.

The approach to volume and calorie estimation can vary based on project requirements. One strategy involves employing separate algorithms for each step: a dedicated volume estimation algorithm for assessing the volume of individual food items within an image, and a distinct calorie estimation algorithm that predicts caloric content based on factors such as food type and nutritional databases. Alternatively, an integrated algorithm can be devised to estimate both volume and calories simultaneously, leveraging ML models that directly associate visual features with caloric values. The choice between these approaches hinges on factors like dataset characteristics, desired accuracy levels, and the inherent relationship between volume and caloric content in the context of the specific project goals. Thorough experimentation and evaluation are essential to determine the most effective strategy for a given use case.

## 5.3 SIGNIFICANCE

The intersection of food and technology is undergoing a transformative shift, driven by advancements in ML. Food image detection and calorie estimation, once a futuristic concept, is now a tangible reality with the potential to revolutionize how we understand and interact with our food. This technology offers a compelling value proposition for individuals, healthcare professionals, and the food industry, paving the way for a healthier and more sustainable future.

**For individuals**

- *Empowered Dietary Choices*: This technology empowers individuals to make informed dietary choices, actively manage weight, and personalize nutrition plans based on their specific needs and goals.
- *Enhanced Food Awareness*: Gaining insights into the macronutrient and micronutrient composition of meals becomes effortless. This newfound awareness empowers individuals to identify potential allergens and dietary restrictions, promoting safer and healthier food choices.
- *Reduced Food Waste*: Food image analysis provides valuable data on portion sizes and consumption patterns, enabling individuals to optimize their food choices and minimize waste.

**For healthcare professionals**

- *Personalized Patient Care*: Nutrition counseling and weight management programs can be significantly enhanced by providing patients with objective data on their food choices and calorie intake. This data empowers healthcare professionals to tailor interventions and treatment plans to individual needs.
- *Early Disease Prevention*: By analyzing food intake patterns, healthcare professionals can identify individuals at risk of diet-related chronic diseases like obesity, diabetes, and heart disease, allowing for early intervention and preventive measures.
- *Improved Patient Outcomes*: With accurate calorie and nutrient estimation, healthcare professionals can monitor patient progress and adjust treatment plans more effectively, ultimately leading to improved patient outcomes.

**For the food industry**

- *Data-Driven Product Development*: Food companies can leverage consumer food choices and preferences gleaned from image analysis to develop innovative products that cater to specific dietary needs and trends.
- *Reduced Food Waste in the Supply Chain*: By optimizing portion sizes and packaging based on real-world consumption data, the food industry can minimize food waste throughout the supply chain, contributing to a more sustainable food system.

## 5.4 CHALLENGES

Implementing a food image detection and calorie estimation project comes with its fair share of challenges, which can be broadly categorized into two main areas.

**Technical challenges**

- *Data Acquisition and Annotation*: Training accurate ML models requires a vast dataset of labeled food images. Acquiring and annotating such a dataset can be time-consuming, expensive, and prone to human error.
- *Image Variability and Complexity*: Food images exhibit immense diversity in terms of lighting, angles, composition, and presentation. This variability makes it challenging for models to generalize accurately and consistently.
- *Mixed Food Items*: When multiple food items (Shimoda W. & Yanai K., 2016) overlap or touch in an image, accurately segmenting and estimating calories for each becomes a complex task.
- *Volume Estimation*: Volume estimation requires to convert the 2D images to 3D images and then calculate the expected volume, which is a complicated process.

**Algorithmic challenges**

- *Model Accuracy and Generalizability*: Developing ML models that achieve high accuracy across a wide range of food types and cuisines remains a challenge. Overfitting to specific datasets (Pouladzadeh., et al., 2015) can lead to poor performance on unseen data.
- *Real-Time Processing Efficiency*: Implementing the technology in real-time applications necessitates efficient algorithms that can analyze images and estimate calories quickly and accurately on mobile devices. There is always a tradeoff between the computational time and model accuracy; maintaining this proper balance can produce an efficient model.

- *Accounting for Nutritional Variations*: The calorie content of food can vary depending on cooking methods, ingredients, and portion sizes. Accurately incorporating these variations into the estimation process remains a hurdle.

Addressing these challenges requires a multidisciplinary approach, combining expertise in computer vision, ML (Chen X., et al., 2017), nutrition science, and human-computer interaction. Continuous research and development efforts, coupled with responsible implementation and community engagement, are key to unlocking the full potential of food image detection and calorie estimation for a healthier and more sustainable future.

## 5.5 SCOPE AND OBJECTIVES

Food recognition plays a pivotal role across diverse domains, ranging from healthcare to personalized nutrition. Understanding its significance involves recognizing its potential in facilitating dietary monitoring, promoting healthier lifestyles, and offering personalized nutrition advice. However, the challenges are notable, with variability in appearance, contextual factors, and environmental conditions posing hurdles to accurate food recognition.

### 5.5.1 Computer vision techniques

The utilization of computer vision techniques, notably CNNs, is central to the success of food recognition systems. This section delves into the effectiveness of these techniques in identifying food items from images. Additionally, it provides insights into crucial aspects like image preprocessing, feature extraction, and other methodologies essential for enhancing the accuracy of food recognition systems.

### 5.5.2 Calorie prediction models

An intrinsic link exists between food recognition and calorie prediction. This section emphasizes the critical role of accurate caloric estimation, particularly for health-conscious individuals. The exploration encompasses various ML models, regression techniques, and data-driven approaches employed to predict caloric content based on recognized food items.

### 5.5.3 Datasets and challenges

The availability of robust datasets is pivotal for training and evaluating food recognition and calorie prediction models. This section reviews existing datasets, discussing their strengths and limitations. Additionally, it addresses the challenges associated with crafting diverse and representative datasets, considering factors such as cultural variations in dietary habits.

### 5.5.4 Applications in health and nutrition

Real-world applications of food recognition and calorie prediction extend beyond theoretical concepts. This section highlights their practical utility in promoting healthy lifestyles, managing dietary conditions, and assisting individuals in achieving fitness goals. The discussion also explores the potential impact of these technologies on healthcare, preventive medicine, and nutrition counseling.

### 5.5.5 Future directions and ethical considerations

The evolving landscape of food recognition and calorie prediction opens up avenues for future exploration. This section delves into emerging trends, such as the integration of wearable devices and advancements in sensor technologies. Simultaneously (Tan M., & Le Q., 2019), it addresses ethical considerations, including privacy concerns and the responsible use of data in the development and deployment of food recognition and calorie prediction systems. This ensures a forward-looking perspective that considers both technological advancements and ethical implications.

## 5.6 EXISTING SOLUTIONS

The burgeoning field of food image recognition and calorie estimation leverages a symphony of sophisticated algorithms to orchestrate a remarkable feat: transforming a simple picture of your meal into an understanding of its nutritional content. Here are some existing solutions related to this project in each phase:

1. *Identifying the Delicacy in the Details (Object Detection)*: At the forefront stands object detection, where algorithms like You Only Look Once (YOLO), faster-Region-based Convolutional Neural Network (RCNN), and Single-Stage object Detection (SSD) meticulously analyze the visual tapestry of your image. Armed with an arsenal of computational tools, they discern the nuances of color, texture, and shape, pinpointing individual food items with an accuracy approaching 90%. Imagine these algorithms as seasoned culinary sleuths, expertly identifying the ingredients composing your culinary puzzle.
2. *Isolating the Dish from the Stage (Image Segmentation)*: Image segmentation algorithms like U-Net and mask R-CNN take the baton. With deft precision, they separate the delectable protagonists – the food items – from the supporting cast – the background. Think of it as drawing a

meticulous curtain around your meal, allowing for a focused analysis of its features. This isolation not only enhances recognition accuracy but also provides valuable data for subsequent calorie estimation.

3. *Gauging the Portion's Grandiose (Volume Estimation)*: However, understanding the complete nutritional picture requires not only knowing the food but also its quantity. Here, portion size estimation algorithms, like those employing image scaling and reference objects, take the center stage. Think of them as skilled stagehands carefully calculating the dimensions of your culinary performance. While their accuracy varies depending on the complexity of the meal, these algorithms typically provide a reliable, 60%–80% accurate, estimate of portion size, completing the nutritional portrait.
4. *Nutrition Estimation*: Once the food items are identified, isolated, and quantized, the spotlight shifts to calorie estimation models. Trained on vast datasets like FOODD100, these DL maestros leverage their knowledge of similar foods and their nutritional profiles to predict the calorie content of your meal. Imagine them as nutritional sages, drawing upon a wealth of culinary wisdom to estimate the energy potential hidden within your dish. With an average accuracy of 75%–85%, these models continually refine their repertoire, striving for ever-more precise estimations.

It's important to acknowledge that this algorithmic orchestra is still under development, constantly refining its performance and addressing concerns regarding accuracy, bias, and user privacy. Yet, the capabilities and potential of these technologies are undeniable, offering individuals the power to make informed dietary choices through the lens of their smartphone camera.

## 5.7 WHY CHOOSE DL

Choosing ML and DL for your food image detection and calorie estimation project holds several compelling advantages over traditional approaches:

**Superior Accuracy and Adaptability**

- Traditional methods like calorie charts and food recognition algorithms often struggle with the immense variability in food appearance in images. ML/DL models, however, can learn complex patterns and relationships from large datasets, enabling them to handle diverse lighting, compositions, and food types with greater accuracy.
- Unlike static rules, ML/DL models can continuously adapt and improve as they are exposed to new data, making them more robust and versatile in dealing with unseen food items and culinary styles.

**Automated Feature Extraction and Segmentation**

- Traditional methods rely on manually defining features like color or texture for food recognition. This process is time-consuming and prone to error. ML/DL algorithms, on the other hand, automatically learn and extract relevant features from the data, simplifying the development process and improving accuracy.
- Image segmentation becomes crucial for identifying and estimating calories of multiple food items in a single image. ML/DL techniques like CNNs excel at this task, accurately separating different food components even when they overlap or touch.

**Real-Time Application Potential**

- With advancements in hardware and efficient network architectures, ML/DL models can be optimized for real-time applications. This opens up exciting possibilities for mobile apps (Kayikci S., 2019) that instantly analyze food images and provide calorie estimates, empowering users to make informed dietary choices on the go.

**Personalization and Dietary Preferences**

- ML/DL models can incorporate individual dietary preferences and restrictions into the calorie estimation process. This allows for personalized calorie tracking and meal planning, catering to specific needs like allergies, vegetarian/vegan diets, or macro/micronutrient goals.

**Scalability and Future Potential**

- ML/DL models are inherently scalable, able to handle increasingly large datasets of food images as they become available. This continuous learning will further refine their accuracy and adaptability, paving the way for even more sophisticated food analysis and nutritional guidance in the future.

## 5.8 ARCHITECTURE INVOLVED

The prerequisite for all these processes to perform is acquiring the dataset. Figure 5.1 is the actual flow of execution of the entire project.

The journey begins with the acquisition of food images, typically captured through smartphones or other imaging devices. These raw images undergo preprocessing steps, including resizing, normalization, and color adjustments, to ensure uniformity and enhance the performance of subsequent algorithms.

To accurately identify individual food items within the images, a segmentation step is employed. This process separates the food items from the background, employing techniques such as graph cut, GrabCut, or advanced

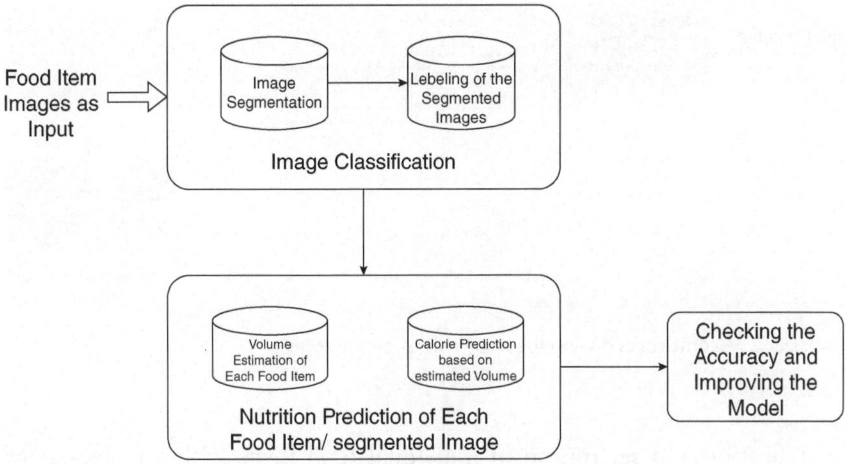


*Figure 5.1* Flow of execution of project.

ML-based segmentation methods such as FCN. But the ML methods produce better results since the process of identifying food items is automated. The precision of segmentation is crucial for downstream tasks.

Once the food items are segmented, the project delves into estimating their volumes. Techniques like stereovision, depth sensors, or volumetric reconstruction from multiple views may be employed. The goal is to create a three-dimensional representation of the food items, allowing for precise volume calculations.

Simultaneously, the system engages in food classification, determining the type of each identified food item. CNNs are commonly used for this purpose, leveraging their capacity to learn hierarchical features directly from images. The accuracy of this step contributes to the reliability of the entire system.

With both volume and food type known, the project moves on to calorie estimation. This can be achieved through various methods, including direct prediction using ML models like CNNs, regression models that analyze visual features, or integration with nutritional databases for precise caloric information retrieval (Farinella G. M., et al., 2016).

The final step involves evaluating the performance of the entire system. Metrics such as accuracy, precision, and recall are assessed. Continuous refinement is crucial, involving feedback loops for model improvement and adjustments to enhance the overall accuracy of the food image recognition and calorie estimation framework as in Figure 5.2.

In the process of food image recognition and calorie estimation, a seamless integration of advanced methods is employed at each step to achieve accurate and reliable results.

At the forefront of image segmentation, the FCN is utilized. FCNs, designed for semantic segmentation, provide pixel-level predictions, ensuring precise

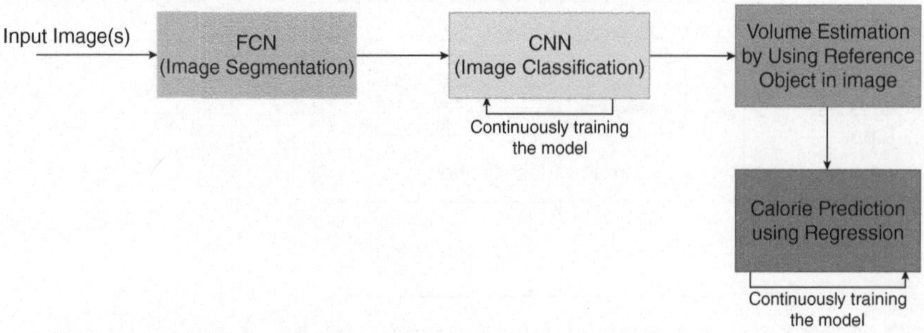


*Figure 5.2* Architecture overview of the entire project.

identification and separation of individual food items within images. This step lays the foundation for subsequent analyses.

For food item classification, the CNN takes the center stage. Leveraging its ability to learn hierarchical features directly from images, the CNN model excels in distinguishing between various food categories. Trained on diverse datasets, it significantly contributes to the accurate classification of different food types.

Volume estimation is a critical step in understanding the spatial dimensions of food items. Here, the reference object method is implemented. By strategically placing reference objects with known dimensions in the images, such as standard plates or containers, the system accurately estimates volumes. This method enables precise 3D reconstructions, enhancing the overall accuracy of volume assessments.

Caloric content prediction is accomplished through a regression model. Trained on visual features extracted from food images, including volume, type, and other relevant parameters, the regression analysis allows for accurate prediction of calorie values. Careful tuning ensures that the model provides reliable estimates of the nutritional content of identified food items.

These integrated methods collectively contribute to a comprehensive and effective approach for extracting detailed nutritional information from food images, empowering users to make informed dietary choices based on accurate estimations of volume and caloric content.

## 5.9 IMPLEMENTATION

### 5.9.1 Data collection

In the pursuit of building a robust food image recognition and nutrition estimation model, the selection of high-quality datasets is paramount. The following are few datasets that can be used to implement this project.

1. *Food24 Dataset*
   The Food24 dataset serves as a foundational component for our image recognition model. It comprises a diverse collection of images spanning a wide array of cuisines and dishes. Collected from various online sources, Food24 provides a rich and varied set of food images, ensuring the model's ability to generalize across different culinary traditions.
2. *ETHz Food 101 Dataset*
   The ETHz Food 101 dataset plays a crucial role in fine-tuning our image recognition model. Focused specifically on food items, this dataset offers a detailed and fine-grained collection of images, each associated with a specific class label. The dataset's granularity allows for a more nuanced understanding of individual food items.
3. *UPMC Food-101 Dataset*
   The UPMC Food-101 dataset further enriches our training data with its extensive collection of food images. Comprising over a hundred thousand images across various classes, this dataset contributes to the model's ability to recognize and classify a broad spectrum of foods.

### 5.9.2 Pre-processing

Data preprocessing is an essential phase in the pipeline of developing a robust food image recognition and nutrition estimation model. It involves transforming raw data into a format suitable for training ML models, enhancing the model's ability to learn patterns and make accurate predictions. The following are common data preprocessing steps.

1. *Image Resizing*:
   - Resizes images to a consistent resolution for uniform input to the model.
   - Common sizes include 224 × 224 or 256 × 256 pixels for CNNs.
2. *Normalization*:
   - Normalizes pixel values to a standard range (e.g., [0, 1] or [–1, 1]).
   - Helps the model converge faster during training.
3. *Missing Data Handling*:
   - Addresses missing data, especially if dealing with additional nutritional information.
   - Imputes or removes missing values based on the dataset size and nature.
4. *Data Augmentation*:
   - Applies augmentation techniques like rotation, flipping, and zooming to artificially increase the diversity of the training dataset.
   - Helps improve the model's ability to generalize to unseen data.

5. *Calorie Estimation*:
   - Ensures accurate mapping of images to corresponding calorie values (Fang S., et al., 2018).
   - Considers incorporating additional features like portion size or ingredient composition for more precise estimations.
6. *Balanced Dataset*:
   - Ensures a balanced distribution of food categories in the dataset.
   - Helps prevent bias and improves the model's ability to recognize different types of food.
7. *Quality Check*:
   - Reviews and validates the dataset for errors, outliers, or inconsistencies.
   - Removes any irrelevant or noisy images that might hinder model performance.

### 5.9.3 Food image classification

Food image classification in food image recognition and calorie prediction involves the use of advanced ML techniques to identify and categorize different food items within images. The goal is to train a model that can accurately recognize a diverse range of foods, enabling subsequent applications such as nutritional analysis and calorie prediction. FCN is one of the best techniques for image classification.

FCNs are a type of neural network architecture that excels in understanding and recognizing patterns in images. When applied to food image classification, FCNs can be employed to distinguish between different types of food items in a given image. Unlike traditional methods that focus on recognizing entire images, FCNs operate at a more detailed level, considering individual pixels. This unique capability enables the network to capture fine-grained information, making it well suited for tasks like recognizing specific features of various foods.

The key strength of FCNs lies in their ability to retain spatial information throughout the network. In a simplified manner, you can think of FCNs as breaking down an image into smaller regions and analyzing each region separately. By doing so, they can understand the distinct characteristics of different foods, even when they appear in varying shapes and sizes. This approach allows FCNs to achieve precise and detailed classifications, making them particularly effective for recognizing and categorizing diverse food items in images.

### 5.9.4 Volume estimation

Volume estimation in food image recognition is a pivotal step for accurate calorie prediction, and a creative approach involves leveraging reference objects, such as a thumb, within the images. This technique relies on the principle that certain reference objects have consistent sizes, enabling the establishment of a scale for volume estimation. In the context of food, a user's thumb can serve as a reliable reference object due to its relatively constant size across individuals. By incorporating the user's thumb as a reference in the image, the system gains a standardized metric that can be used to infer the dimensions of the food items, allowing for an estimation of their volumes.

To implement this technique, the system first identifies and extracts the reference object, i.e., the thumb, within the image using image processing and computer vision algorithms. Once the thumb is localized, its dimensions are utilized to create a scale factor that can be applied to estimate the size of other food items in the image. The identified thumb provides a consistent reference point, overcoming potential variations in camera angles and distances. This approach not only aids in volume estimation but also enhances the system's ability to accurately predict the caloric content of the food by incorporating spatial information about its size.

By employing reference objects like the thumb for volume estimation, the system becomes more versatile and adaptable to diverse images, catering to varying perspectives and distances from the camera. This technique exemplifies the innovative solutions within the realm of food image recognition and calorie prediction, demonstrating how leveraging common objects within images can significantly enhance the accuracy of volume estimation and, consequently, the precision of nutritional predictions in dietary analysis applications.

### 5.9.5 Nutrition estimation

Nutrition estimation, the final step in food image recognition and calorie prediction, involves associating recognized food items with their respective nutritional information. After successfully identifying and classifying the food items within an image, the system refers to a comprehensive nutrition database containing details such as macronutrient content, vitamins, minerals, and other relevant nutritional values for various foods. Each recognized food class is matched with the corresponding entry in the nutrition database to retrieve its nutritional profile.

Once the matching process is complete, the system aggregates the nutritional information of all identified food items to provide an overall assessment of the meal's nutritional content. This includes calculating the total calories, proteins, fats, carbohydrates, and other key nutritional components. Advanced ML models may enhance the precision of this estimation by considering factors such as portion sizes, cooking methods, and ingredient combinations. The result is a holistic nutritional analysis that empowers users to make informed decisions about their dietary choices based on the visual information extracted from the food images.

In summary, nutrition estimation involves a symbiotic relationship between food image recognition, calorie prediction, and a robust nutrition database. By leveraging ML and referencing a comprehensive nutritional repository, the system can deliver valuable insights into the dietary composition of a given meal, contributing to a more informed and personalized approach to nutrition management.

## 5.10 RESULTS

In the study's results, the utilization of FCN for food image classification showcased promising outcomes, demonstrating a high degree of accuracy in identifying various food items. Comparative analysis against existing models highlighted the effectiveness of the FCN approach, revealing its potential in enhancing the classification accuracy. Furthermore, the integration of a thumb as a reference object for food volume estimation yielded reliable results, validating its practicality in estimating the quantity of depicted food items. Robustness tests across diverse food types indicated the model's ability to generalize well, contributing to its versatility in real-world applications.

The final step of nutrition estimation, encompassing the classified food images and estimated volumes, produced accurate predictions of calorie content without delving into specific technical details. Real-world case studies illustrated instances where the entire process, from image classification to nutrition estimation, effectively reflected the true nutritional content of various meals. While certain challenges and limitations were encountered, particularly in scenarios with complex food compositions, user feedback confirmed the system's overall reliability and practical utility. In conclusion, the study's findings contribute valuable insights into the potential of this approach for non-technical readers seeking an understanding of the capabilities and limitations of food image recognition and nutrition estimation systems.

## 5.11 CONCLUSIONS AND RESEARCH DIRECTIONS

In conclusion, the project on food image recognition and calorie estimation has showcased substantial potential in addressing global health concerns, particularly obesity and overeating. Leveraging DL techniques has not only bolstered accuracy but has also led to reduced response times and lowered energy consumption, marking a significant advancement in dietary management.

The success of the project can be attributed to a comprehensive layer-based approach encompassing image acquisition, food item classification, surface area detection, and calorie prediction. This methodology has proven effective in transforming our approach to dietary monitoring, laying the foundation for more intuitive and user-friendly applications that facilitate informed dietary choices. While the current results are promising, there exists ample opportunity for improvement and optimization. Future endeavors may explore sophisticated models and larger, more diverse datasets to further enhance system performance, ensuring its robustness and reliability.

This project serves as a testament to the transformative power of ML and image processing in reshaping how we manage our diets and cultivate health consciousness. It signifies a crucial step toward a future where technology becomes a powerful ally, aiding individuals in tracking their food intake and fostering healthier lifestyles for improved overall health outcomes.

## REFERENCES

Chen X., Zhu Y., Zhou H., Diao L., & Wang D., "ChineseFoodNet: A large-scale image dataset for Chinese food recognition," 2017, arXiv:1705.02743.

Chopra M. & Purwar A., "Food image recognition by optimizing CNN with PSO and GA," in *Proc. 14th Int. Conf. Contemp. Comput.*, New York, NY, USA, Aug. 2022, pp. 37–42.

Chotwanvirat P., Hnoohom N., Rojroongwasinkul N. & W. Kriengsinyos, "Feasibility study of an automated carbohydrate estimation system using Thai food images in comparison with estimation by dietitians," *Frontiers Nutrition*, vol. 8, Oct. 2021, Art. no. 732449.

Fakhrou A., Kunhoth J., & Al Maadeed S., "Smartphone-based food recognition system using multiple deep CNN models," *Multimedia Tools Appl.*, vol. 80, nos. 21–23, pp. 33011–33032, Sep. 2021.

Fang S., Shao Z., Mao R., Fu C., Delp E. J., Zhu F., Kerr D. A. & Boushey C. J., "Single-view food portion estimation: Learning image-toenergy mappings using generative adversarial networks," in *Proc. 25th IEEE Int. Conf. Image Process. (ICIP)*, Oct. 2018, pp. 251–255.

Farinella G. M., Allegra D., Moltisanti M., Stanco F., & Battiato S., "Retrieval and classification of food images," *Comput. Biol. Med.*, vol. 77, pp. 23–39, Oct. 2016.

Gao J., Tan W., Ma L., Wang Y., & Tang W., "MUSEFood: Multisensor-based food volume estimation on smartphones," in *Proc. IEEE SmartWorld, Ubiquitous Intell. Comput., Adv. Trusted Comput., Scalable Comput. Commun., Cloud Big Data Comput., Internet People Smart City Innov. (SmartWorld/SCALCOM/UIC/ATC/CBDCom/IOP/SCI)*, Aug. 2019, pp. 899–906.

He Y., Xu C., Khanna N., Boushey C. J., & Delp E. J., "Food image analysis: Segmentation, identification and weight estimation," in *Proc. IEEE Int. Conf. Multimedia Expo (ICME)*, Jul. 2013, pp. 1–6.

Jiang L., Qiu B., Liu X., Huang C., & Lin K., "DeepFood: Food image analysis and dietary assessment via deep model," *IEEE Access*, vol. 8, pp. 47477–47489, 2020.

Kawano Y. & Yanai K., "FoodCam-256: A large-scale real-time mobile food RecognitionSystem employing high-dimensional features and compression of classifier weights," in *Proc. 22nd ACM Int. Conf. Multimedia*, Nov. 2014, pp. 761–762.

Kayikci S., Basol Y., & Dorter E., "Classification of Turkish cuisine with deep learning on mobile platform," in *Proc. 4th Int. Conf. Comput. Sci. Eng.* (UBMK), 2019, pp. 1–5.

Minija S. J. & Emmanuel W. R. S., "Imperialist competitive algorithm based deep belief network for food recognition and calorie estimation," *Evol. Intell.*, vol. 15, no. 2, pp. 955–970, Jun. 2022.

Pouladzadeh P., Shirmohammadi S., and Al-Maghrabi R., "Measuring calorie and nutrition from food image," *IEEE Trans. Instrum. Meas.*, vol. 63, no. 8, pp. 1947–1956, Aug. 2014.

Pouladzadeh P., Yassine A., & S. Shirmohammadi, "FooDD: Food detection dataset for calorie measurement using food images," in *Proc. Int. Conf. Image Anal. Process*. Cham, Switzerland: Springer, 2015, pp. 441–448.

Shimoda W. & Yanai K., "Foodness proposal for multiple food detection by training of single food images," in *Proc. 2nd Int.* Workshop Multimedia Assist. Dietary Manag., Oct. 2016, pp. 13–21.

Sudo K., Murasaki K., Shimamura, & Taniguchi Y., "Estimating nutritional value from food images based on semantic segmentation," in *Proc. ACM Int. Joint Conf. Pervasive Ubiquitous Comput.*, Adjunct Publication, New York, NY, USA, Sep. 2014, pp. 571–576.

Tan M., & Le Q., "EfficientNet: Rethinking model scaling for convolutional neural networks," in *Proc. Int. Conf. Mach. Learn.*, 2019, pp. 6105–6114.

Chapter 6

# Precision agriculture

## Navigating environmental sustainability

*Vimal Gaur, Shipra Raheja, Rachna Jain and Ranjit Singh Sarban Singh*

### 6.1 INTRODUCTION

Smart crop management involves leveraging technology, such as sensors and data analytics, to optimize factors like irrigation, fertilization, and pest control. By monitoring and adjusting these variables based on real-time data, farmers can improve crop yield and reduce resource wastage. Several factors influence smart crop management for increased yield, including data accuracy, technology integration, climate conditions, etc. Reliable data from sensors and other sources is crucial for making informed decisions in areas like irrigation and fertilization. Successful smart crop management requires seamless integration of technologies like Internet of Things (IoT), artificial intelligence (AI), and data analytics to automate and optimize processes. Adapting strategies to changing weather patterns and seasonal variations is essential for maximizing yield. Different crops have distinct requirements, so tailoring management practices to specific varieties ensures optimal result [1]. Efficient use of resources such as water and nutrients helps minimize waste and environmental impact while maximizing productivity [2]. Early detection and targeted responses to pests and diseases are critical to preventing yield losses. The scale and layout of the farm impact the deployment of technology and the overall efficiency of management practices. Ensuring that farmers are educated and trained in the use of smart technologies is vital for successful implementation. Adequate infrastructure, including connectivity and power supply, is necessary for the functioning of smart systems in remote agricultural areas. The cost-effectiveness of implementing smart crop management practices influences their widespread adoption among farmers. Addressing these factors collectively contributes to the success of smart crop management systems in enhancing agricultural yield.

### 6.2 SMART CROP MANAGEMENT KEY ELEMENTS

"Smart" in smart crop management refers to the integration of advanced technologies and data-driven approaches to enhance agricultural practices.

DOI: 10.1201/9781003628392-6

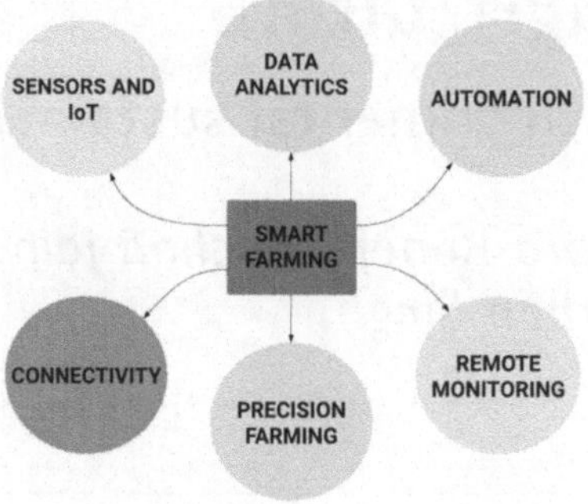


*Figure 6.1* Smart farming key elements.

The key elements include sensors and IoT, data analytics, precision farming, etc., as shown in Figure 6.1.

1. Sensors and IoT: Deploying sensors to collect real-time data on soil moisture, temperature, and other variables. IoT devices facilitate connectivity and communication between different components [3].
2. Data Analytics: Employing data analytics and AI to analyze the collected information, providing insights for informed decision-making.
3. Automation: Using automated systems and robotics for tasks such as planting, harvesting, and monitoring, reducing labor requirements and improving efficiency.
4. Precision Farming: Implementing precise and targeted applications of inputs, such as water, fertilizers, and pesticides, based on data analysis and specific crop needs [4].
5. Connectivity: Utilizing interconnected systems that enable seamless communication between devices, machines, and databases, facilitating coordinated and efficient operations.
6. Remote Monitoring: Enabling farmers to monitor and manage their crops remotely through technology, allowing timely responses to changing conditions.
7. Adaptive Strategies: Adjusting agricultural practices dynamically based on real-time data, weather forecasts, and other factors to optimize crop growth and resource use.

The "smart" aspect involves leveraging these technologies to create a more efficient, sustainable, and productive approach to crop management [5].

## 6.3 MECHANISMS OF POSITIVE IMPACT IN SMART CROP MANAGEMENT

Smart crop management involves the application of advanced technologies and data-driven approaches to optimize agricultural practices. This includes utilizing sensors, IoT devices, AI, and data analytics to monitor and control various factors influencing crop growth. By integrating these technologies, farmers can make informed decisions on irrigation, fertilization, pest control, and other crucial aspects, ultimately leading to increased yield, resource efficiency, and sustainability in agriculture. Smart crop management positively impacts yield through several mechanisms:

1. Precision Farming: Technologies like Global Positioning System (GPS)-guided tractors and drones enable precise planting, harvesting, and application of inputs, reducing wastage and optimizing crop growth.
2. Data-Driven Decision-Making: Real-time data from sensors helps farmers make informed decisions on irrigation, fertilization, and pest control, adapting to changing conditions promptly.
3. Optimized Resource Use: Smart systems efficiently manage resources like water and nutrients, minimizing waste and ensuring that crops receive the right amount at the right time.
4. Early Pest and Disease Detection: Monitoring systems can identify potential issues before they become widespread, allowing farmers to take targeted actions, preventing significant yield losses.
5. Climate Adaptation: Smart crop management systems can adjust strategies based on weather patterns, helping crops cope with changing environmental conditions.
6. Automation: Automated processes, such as robotic harvesting, can improve efficiency and reduce labor costs, contributing to increased overall yield [6].
7. Crop Variety Optimization: Tailoring management practices to specific crop varieties ensures that each type receives the necessary care for optimal growth and productivity.

By integrating these elements, smart crop management enhances overall agricultural efficiency, reduces resource input, and mitigates risks, ultimately leading to increased crop yields.

## 6.4 SMART CROP MANAGEMENT: NAVIGATING ENVIRONMENTAL IMPACTS

The implementation of smart crop management can bring about several positive environmental changes:

1. Resource Conservation: Smart systems optimize resource use, reducing water, fertilizers, and pesticides, minimizing environmental impact, and preserving vital resources.
2. Reduced Chemical Usage: Precision application of inputs based on data-driven insights helps minimize the need for broad-spectrum chemicals, promoting healthier ecosystems.

3. Erosion Prevention: Smart farming practices, such as controlled irrigation, can prevent soil erosion, preserving soil structure and fertility, which is crucial for long-term environmental health.
4. Biodiversity Support: Precision agriculture allows for more targeted interventions, reducing the negative impact on surrounding biodiversity by minimizing disruption to non-target species.
5. Lower Greenhouse Gas Emissions: Automation and optimized processes can lead to reduced energy consumption and, consequently, lower greenhouse gas emissions, contributing to climate change mitigation.
6. Water Quality Improvement: By reducing runoff from excess irrigation or the use of chemicals, smart crop management can contribute to better water quality in nearby water bodies.
7. Climate Resilience: Adaptive strategies based on real-time data can help crops better cope with changing weather patterns, promoting resilience to climate change.
8. Sustainable Land Use: Smart crop management supports efficient land use, ensuring that agricultural practices are sustainable without causing undue stress on ecosystems.

While the positive environmental changes are substantial, it's essential to note that the successful integration of smart crop management should be accompanied by responsible and sustainable farming practices to maximize these benefits.

## 6.5 MAXIMIZING YIELD: THE MULTIFACETED IMPACT OF SMART CROP MANAGEMENT

Smart crop management represents a paradigm shift in agriculture, orchestrating a symphony of technologies and strategies to optimize crop yield. This holistic approach encompasses various mechanisms, each contributing to the overall enhancement of agricultural productivity. Smart crop management positively impacts yield through several mechanisms, viz. optimizing resource uses especially optimizing variety of crops. Further early pest and disease detection, and adaption according to climate should also be done, On integration of optimized resources, and automated processes smart crop management can be done efficiently. By integrating these elements, smart crop management enhances overall agricultural efficiency, reduces resource input, and mitigates risks, ultimately leading to increased crop yields.

## 6.6 SMART AGRICULTURE AND TECHNIQUES

Smart agriculture employs various techniques that leverage technology to enhance efficiency, sustainability, and productivity in farming. Some key

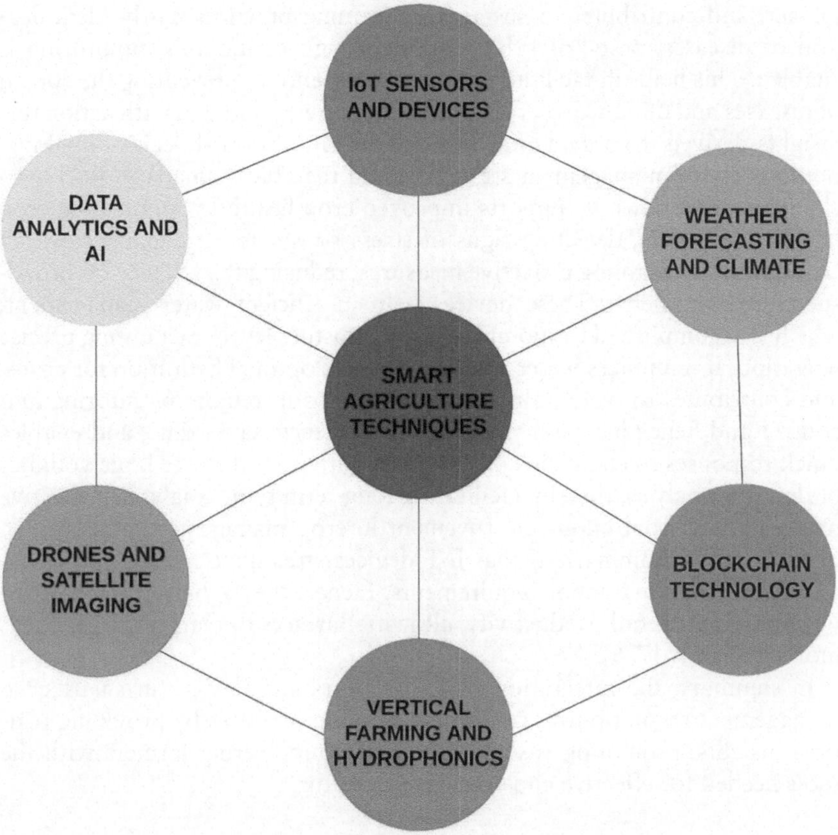


*Figure 6.2* Smart crop management techniques.

techniques are shown in Figure 6.2, and their impact has been discussed in this section.

## 6.6.1 IoT sensors and devices

The prime impact of IoT sensors and devices is real-time monitoring, which is providing continuous, real-time data on crucial factors such as soil moisture, temperature, and nutrient levels. It enables farmers to make prompt and informed decisions, optimizing resource use and ensuring optimal crop conditions. IoT sensors and devices facilitate precision farming by delivering accurate and detailed information about specific areas within a field. It allows for targeted interventions, minimizing input wastage, and maximizing the efficiency of crop management practices. These devices also lead to resource optimization, which is enhancing resource efficiency by monitoring and controlling irrigation, fertilization, and pesticide application based on actual needs. It also reduces the overuse of resources, minimizes environmental impact, and contributes to sustainable farming practices. Early identification of diseases, pests, or other issues through continuous monitoring is enabled. This helps in facilitating timely intervention, preventing the spread of diseases and minimizing crop losses. It empowers farmers with actionable insights derived from data analytics and aids in informed decision-making, optimized crop management strategies, and the ability to adapt quickly to changing conditions. It supports improved crop health by monitoring crop health parameters, detecting signs of stress or disease. It enhances overall crop health by enabling proactive measures, reducing the reliance on broad-spectrum treatments. These devices help in efficient water management, which is accomplished by monitoring soil moisture levels, facilitating precise irrigation. It minimizes water wastage, ensures optimal hydration for crops, and contributes to water conservation. It helps in remote monitoring and control and hence increases operational efficiency, saves time, and enables quick responses to changing conditions. It gathers and stores large volumes of data for analysis, thereby facilitating long-term trend analysis, predictive modeling, and continuous improvement in crop management practices. In brief, it can be summarized that IoT devices streamline farming processes and reduces manual labor requirements. Hence, they achieve the objective of improving overall productivity, allowing farmers to manage larger areas more efficiently [7, 8].

In summary, the integration of IoT sensors and devices in smart crop management revolutionizes traditional farming practices by providing real-time insights, promoting sustainability, and empowering farmers with the tools needed for effective and efficient agriculture.

### 6.6.2 Precise Farming

Precision farming, an integral component of smart crop management, is a revolutionary approach to agriculture that leverages cutting-edge technologies to enhance efficiency, productivity, and sustainability. This method goes beyond traditional farming practices by integrating data, sensors, and advanced analytics into the decision-making process. At the heart of precise farming is the utilization of various technologies to gather real-time data about crop conditions. Sensors placed in fields monitor crucial parameters such as soil moisture, temperature, and nutrient levels. Drones equipped with multispectral cameras capture high-resolution images, providing farmers with detailed insights into crop health and identifying potential issues like diseases or pest infestations. This data forms the foundation for informed decision-making, allowing farmers to tailor their actions to specific needs. One of the key aspects of precise farming is targeted resource application. Instead of uniformly applying water, fertilizers, or pesticides across an entire field, farmers can use data-driven insights to apply these resources precisely where they are needed. This not only conserves resources but also minimizes environmental impact. For example, through variable rate technology, automated systems can adjust the amount of irrigation or fertilizers based on the specific requirements of different areas within a field. Furthermore, smart crop management employs automation to streamline farming operations. Autonomous machinery, guided by GPS and data inputs, can perform tasks such as planting, harvesting, and weed control with unparalleled accuracy. This not only reduces labor costs but also ensures that each plant is treated optimally, contributing to increased overall yield. The integration of data analytics is another crucial element of precise farming. Advanced algorithms process the vast amounts of data collected from sensors and other sources to generate actionable insights. These insights can include predictions about crop yield, identification of potential issues, and recommendations for optimal planting times. By harnessing the power of data, farmers can make strategic decisions that significantly impact the success of their crops. In addition to enhancing productivity, precise farming plays a pivotal role in sustainable agriculture. By minimizing the use of resources and reducing environmental impact, farmers can contribute to long-term ecological balance. The ability to monitor and manage crops with precision enables a more efficient use of water, fertilizers, and pesticides, aligning agricultural practices with environmental conservation goals. In conclusion, precise farming within the framework of smart crop management represents a transformative approach to agriculture. By leveraging technology to gather, analyze, and apply data, farmers can optimize their operations, increase yields, and contribute to sustainable and environmentally conscious farming practices. As technology continues to evolve, the potential for precision farming to shape the future of agriculture is vast and promising.

### 6.6.3 Weather forecasting and climate

Weather forecasting and climate modeling are crucial components of smart crop management, providing farmers with valuable insights to make informed decisions about planting, harvesting, and overall crop care. Weather forecasting involves predicting short-term atmospheric conditions, including temperature, precipitation, and wind patterns. Accurate and timely weather forecasts enable farmers to plan their activities more effectively. For example, knowing when rain is expected allows for optimized irrigation scheduling, while awareness of upcoming temperature changes can influence decisions regarding frost protection or heat stress management. Climate modeling, on the other hand, focuses on long-term patterns and trends. By analyzing historical climate data and utilizing sophisticated models, scientists can make projections about future climate conditions. This information is invaluable for farmers planning crops with longer growing cycles, helping them adapt to changing climate patterns and make informed decisions about the types of crops to cultivate. In smart crop management, the integration of weather forecasting and climate modeling is essential for precision agriculture. Farmers can use this information to adjust their planting and harvesting schedules, optimize irrigation practices, and implement targeted pest control measures. For instance, if a weather forecast indicates an increased risk of disease due to specific climatic conditions, farmers can proactively apply preventive measures. Advanced technologies, such as sensors and satellite imagery, complement weather forecasting and climate modeling by providing real-time data for specific locations. This localized information enhances the accuracy of decision-making, allowing farmers to tailor their actions to the unique conditions of their fields. Data analytics, often employed in conjunction with weather and climate data, further refines smart crop management strategies. Machine learning (ML) algorithms can analyze historical weather patterns, crop responses, and yield outcomes to make predictions and recommendations. This predictive capability helps farmers anticipate challenges and optimize their farming practices for better outcomes.

The collective impact of these smart agriculture techniques is a more sustainable, efficient, and resilient agricultural sector. By harnessing technology, farmers can navigate challenges, increase productivity, and contribute to a more environmentally conscious and food-secure future [9, 10].

### 6.6.4 Blockchain technology

Blockchain technology is increasingly being integrated into smart crop management systems, providing transparency, traceability, and security throughout the agricultural supply chain. One key application of blockchain in crop management is supply chain transparency. Each stage of the agricultural process, from planting to harvesting, processing, and distribution, can be recorded as a block in the blockchain. This creates an immutable and transparent ledger of information, allowing stakeholders, including farmers, distributors, and consumers, to trace the journey of crops from the field to the market. This transparency enhances trust and helps in verifying the authenticity of the produce. Smart contracts, which are self-executing contracts with the terms of the agreement directly written into code, can be integrated into blockchain-based crop management systems. These contracts automate various processes, such as payment transactions and quality assurance checks. For instance, when predefined conditions (like the successful delivery of a batch of crops) are met, the smart contract can automatically trigger payments, reducing the need for intermediaries and streamlining transactions. Blockchain also addresses the issue of data integrity in agriculture. As data related to crop conditions, weather patterns, and other factors is recorded on the blockchain, it becomes resistant to tampering or manipulation. This ensures the accuracy and reliability of the information used in decision-making processes [11].

In terms of quality control, blockchain can be utilized to record and track data related to factors like soil conditions, irrigation, and pest control methods. This data can then be used to certify the quality and sustainability of crops. Consumers can access this information, fostering transparency and allowing them to make informed choices about the produce they purchase. Furthermore, blockchain facilitates the creation of decentralized marketplaces, connecting farmers directly with consumers or retailers. This disintermediation can lead to fairer pricing and more direct relationships between producers and buyers. In conclusion, blockchain technology in smart crop management provides a secure, transparent, and efficient framework for managing and recording agricultural data. By enhancing traceability, automating processes through smart contracts, and ensuring data integrity, blockchain contributes to a more reliable and sustainable agricultural supply chain [12].

### 6.6.5 Vertical farming and hydroponics

Vertical farming and hydroponics represent innovative approaches to smart crop management, offering efficient and sustainable solutions for maximizing crop production in limited spaces. Vertical farming involves cultivating crops in vertically stacked layers or vertically inclined surfaces, often in controlled indoor environments [13]. This method optimizes space utilization, making it suitable for urban areas where land availability is limited. Vertical farming systems integrate advanced technologies, such as LED lighting and climate control, to create optimal growing conditions for plants. These controlled environments enable year-round cultivation, independent of external weather conditions. Hydroponics is a soilless cultivation method where plants receive nutrients through a nutrient-rich water solution. This approach eliminates the need for traditional soil, providing greater control over nutrient levels and minimizing the risk of soil-borne diseases. In smart crop management, hydroponic systems often utilize automation and sensors to monitor and adjust factors like nutrient concentrations, pH levels, and water usage. This precision contributes to efficient resource utilization and improved crop yields. The combination of vertical farming and hydroponics enhances the overall efficiency of crop production. Vertical stacking allows for a higher density of crops, effectively multiplying the available growing space [14]. Hydroponic systems ensure that plants receive precisely calibrated nutrients, promoting faster growth and healthier crops. Smart crop management in vertical farming and hydroponics often incorporates data analytics and automation. Sensors monitor environmental variables such as temperature, humidity, and light intensity. Automated systems respond to these data inputs by adjusting conditions to meet the specific requirements of the crops. This data-driven approach enhances precision and allows for optimal crop management. Additionally, the controlled environment of vertical farming and hydroponic systems reduces the reliance on pesticides and minimizes water usage compared to traditional farming methods. The closed-loop systems in hydroponics also contribute to water conservation, as the nutrient solution can be recirculated, reducing waste. In conclusion, vertical farming and hydroponics exemplify smart crop management by leveraging technology to create efficient and sustainable cultivation systems.

These methods address challenges related to land scarcity, resource optimization, and environmental impact, ultimately contributing to more resilient and productive agriculture in both urban and traditional settings.

### 6.6.6 Smart irrigation systems

Smart irrigation systems are integral components of smart crop management, revolutionizing the way farmers manage water resources and optimize crop growth. These systems leverage technology to enhance precision, conserve water, and improve overall agricultural sustainability. One key feature of smart irrigation is real-time monitoring. Soil moisture sensors placed in the field continuously measure moisture levels, allowing farmers to receive instant data about the hydration status of their crops. This information helps in determining when and where irrigation is needed, preventing both over- and under-watering. Over time, this precise approach contributes to water conservation and increased efficiency. Weather forecasting integration is another essential aspect of smart irrigation. By incorporating real-time weather data, these systems can anticipate upcoming rainfall or changes in temperature. This foresight enables automated adjustments to irrigation schedules, ensuring that water is applied based on the actual needs of the crops and reducing wastage. Furthermore, smart irrigation systems often employ drip or precision irrigation techniques. Drip irrigation delivers water directly to the root zone of plants, minimizing evaporation and runoff. This targeted approach not only conserves water but also enhances nutrient uptake by plants, promoting healthier crop growth. Automation is a key feature in smart irrigation, allowing farmers to remotely control and monitor irrigation systems. This can be done through mobile applications or centralized control systems. Automated irrigation schedules, triggered by data inputs from soil sensors or weather forecasts, streamline the irrigation process and reduce the need for constant manual oversight. Data analytics also play a role in optimizing smart irrigation. By analyzing historical data on soil moisture, weather patterns, and crop growth, farmers can fine-tune their irrigation strategies over time. ML algorithms may even predict optimal irrigation schedules, contributing to resource efficiency and improved crop yields. In conclusion, smart irrigation systems are transformative in smart crop management. By combining real-time monitoring, weather integration, precision techniques, automation, and data analytics, these systems empower farmers to make informed decisions about water usage. This not only conserves a critical resource but also contributes to sustainable agriculture practices, ultimately improving the productivity and resilience of crop cultivation.

### 6.6.7 Data analytics and AI

Data analytics and AI play pivotal roles in modern crop management, enabling farmers to make data-driven decisions and optimize agricultural practices [15, 16]. Data analytics processes large sets of information collected from various sources, such as sensors, satellites, and drones. By analyzing this data, farmers gain valuable insights into soil health, weather patterns, crop conditions, and more. This information forms the basis for strategic decision-making, allowing farmers to tailor their actions to specific needs and challenges in real time.

AI algorithms enhance the capabilities of data analytics by extracting patterns, predicting outcomes, and offering actionable recommendations. For example, ML models can analyze historical data to predict optimal planting times, identify potential disease outbreaks, or recommend personalized irrigation schedules. This predictive capability empowers farmers to proactively address issues, optimize resource allocation, and ultimately increase crop yields.

One key application of data analytics and AI in crop management is precision agriculture. By leveraging real-time data, farmers can implement precise interventions, such as targeted irrigation, fertilization, and pest control. This not only maximizes the efficiency of resource use but also minimizes environmental impact, contributing to sustainable farming practices.

In disease management, AI can analyze images of crops captured by drones or cameras to detect early signs of diseases or pest infestations. This early detection allows for timely and targeted responses, reducing the spread of diseases and minimizing crop losses.

Moreover, AI-driven farm management systems can integrate various data sources to create comprehensive field maps. These maps can highlight variations in soil composition, moisture levels, and other factors, enabling farmers to implement site-specific strategies for different areas within a field.

### 6.6.8 Robotics and automation

Robotics and automation are key components of smart crop management, offering farmers advanced tools to increase efficiency, reduce labor costs, and optimize agricultural practices [3, 17]. One significant application of robotics in smart crop management is in planting and harvesting. Autonomous machines equipped with precision technology and sensors can navigate fields, planting seeds with accuracy and harvesting crops at optimal times. These robotic systems not only streamline operations but also contribute to increased yield by ensuring precise and timely actions.

In addition to planting and harvesting, robotics plays a vital role in weed control. Robotic systems can identify and eliminate weeds with precision, reducing the need for herbicides and minimizing damage to crops. This targeted approach helps in maintaining a healthy crop while minimizing environmental impact.

Automation also extends to irrigation and fertilization processes. Smart systems, often integrated with sensors and data analytics, can monitor soil conditions in real time. Based on this information, automated irrigation systems can deliver the right amount of water precisely where it's needed. Similarly, automated fertilization systems can adjust nutrient levels based on the specific requirements of different areas within a field. This level of precision minimizes resource wastage and contributes to sustainable farming practices.

Furthermore, drones equipped with robotic capabilities can be deployed for tasks such as crop monitoring, disease detection, and even precision spraying. These autonomous aerial vehicles can cover large areas quickly, capturing data and implementing targeted interventions efficiently [18].

The integration of robotics and automation in smart crop management leads to a more data-driven and responsive agricultural system. Advanced analytics, combined with the capabilities of robotic systems, enable farmers to make informed decisions in real time. This not only improves overall productivity but also allows for a proactive approach to addressing challenges such as pest infestations or crop diseases [19].

In conclusion, the use of robotics and automation in smart crop management represents a transformative shift in agriculture. These technologies empower farmers to enhance precision, reduce reliance on manual labor, and optimize resource utilization. As technology continues to advance, the potential for further innovations in robotics and automation within agriculture is considerable, promising increased sustainability and efficiency in food production [20].

### 6.6.9 Drones and satellite imaging

Drones and satellite imaging play pivotal roles in smart crop management, providing farmers with valuable insights and enhancing precision in agriculture [21].

Satellite imaging offers a macroscopic view of large agricultural areas. High-resolution satellite imagery can monitor crop health, identify potential issues, and assess overall field conditions. This data, often obtained at regular intervals, allows farmers to track changes over time and make informed decisions regarding irrigation, fertilization, and pest control. Satellite technology provides a broad perspective that aids in strategic planning and optimizing resource allocation.

On the other hand, drones offer a more localized and detailed perspective. Equipped with advanced cameras and sensors, drones can capture high-resolution images of crops, soil, and infrastructure at a much finer scale. This enables farmers to detect early signs of diseases, assess the effectiveness of treatments, and monitor specific areas within a field. Drones are particularly useful in providing real-time data, allowing for swift responses to emerging issues [22].

The combination of satellite imaging and drone technology allows for a comprehensive approach to crop monitoring. Satellite data provides the big picture, while drones offer detailed, on-the-ground insights. Integrating these technologies enables farmers to make precise and timely decisions, optimizing resource use and ultimately improving crop yield.

Smart crop management systems often use data from both satellites and drones to create detailed maps of fields. These maps can highlight variations in soil health, identify areas prone to diseases or pests, and help in creating tailored treatment plans. By leveraging the strengths of both satellite and drone technology, farmers can implement targeted interventions, minimizing the use of resources and maximizing the efficiency of their operations [23].

In conclusion, the integration of drones and satellite imaging in smart crop management revolutionizes the way farmers monitor and manage their fields. These technologies provide a synergistic approach, combining the broad coverage of satellites with the detailed, localized insights from drones. By harnessing the power of aerial data, farmers can enhance precision, optimize resource use, and contribute to sustainable and efficient agriculture practices [24].

## REFERENCES

[1] S. Hafner, "Trends in maize, rice, and wheat yields for 188 nations over the past 40 years: a prevalence of linear growth," *Agric. Ecosyst. Environ.*, vol. 97, no. 1–3, pp. 275–283, 2003.

[2] D.B. Lobell et al, "On the use of statistical models to predict crop yield responses to climate change," *Agric. For. Meteorol.*, vol. 150, no. 11, 1443–1452, 2010.

[3] N. Johnson, M. B. S. Kumar, and T. Dhannia, "A study on the significance of smart IoT sensors and data science in digital agriculture," in *Proc. Adv. Comput. Commun. Technol. High Perform. Appl. (ACCTHPA)*, Cochin, India, Jul. 2020, pp. 80–88.

[4] X. Zhang et al, "Contribution of cultivar, fertilizer and weather to yield variation of winter wheat over three decades: a case study in the North China Plain," *Eur. J. Agron.*, vol. 50, 52–59, 2013.

[5] M.R.M. Kassim, "IoT Applications in Smart Agriculture: Issues and Challenges", In *Proceedings of the IEEE Conference on Open Systems (ICOS)*, pp. 17–19, November 2020.

[6] J. Meng," Research on the cost of agricultural products circulation and its control under the new normal economic development." *Commercial Times*, vol. 12, no. 23, pp. 145147, 2016.

[7] . S. Navulur and M. N. Giri Prasad, "Agricultural management through wireless sensors and Internet of Things," *Int. J. Elect. Comput. Eng*, vol. 7, no. 6, pp. 3492, 2017.

[8] I. M. Marcu, G. Suciu, C. M. Balaceanu and A. Banaru, "IoT based System for Smart Agriculture", *2019 11th International Conference on Electronics Computers and Artificial Intelligence (ECAI)*, pp. 1–4, 2019.

[9] . P. Suebsombut, A. Sekhari, P. Sureepong, P. Ueasangkomsate, and A. Bouras, "The using of bibliometric analysis to classify trends and future directions on 'smart farm,"' in *Proc. Int. Conf. Digit. Arts, Media Technol. (ICDAMT)*, Chiang Mai, Thailand, 2017, pp. 136–141, doi: 10.1109/ICDAMT.2017.7904950

[10] Y. Salima, J. F. M. Peira, G. R. de la Horra, and P. V. M. Ablanque, "Remote sensing data: Useful way for the precision agriculture," in *Proc. 6th Int. Conf. Internet Things, Syst., Manage. Secur. (IOTSMS)*, Granada, Spain, Oct. 2019, pp. 603–609.
[11] R.R. Shamshiri, C. Weltzien, I.A. Hameed, I.J. Yule, T.E. Grift, S.K. Balasundram, L. Pitonakova, D. Ahmad, and G. Chowdhary," Research and development in agricultural robotics: A perspective of digital farming," *Int. J. Agric. Biol. Eng.* 11, 1–14, 2018.
[12] Shibin David, RS Anand and Martin Sagayam, "Enhancing AIBased Evaluation for Smart Cultivation and Crop Testing Using Agro-Datasets," *J. Artif. Intell. Res.*, vol. 2, no. 1, pp. 149–167, 2020.
[13] M. Bacco, A. Berton et al., "IEEE 802.15.4 Air-Ground UAV Communications in Smart Farming Scenarios", *IEEE Commun. Lett*, vol. 22, pp. 1910–1913, 2018.
[14] R. Divya and R Chinnaiyan, "Reliable AI-Based Smart Sensors for Managing Irrigation Resources in Agriculture—A Review", in *International Conference on Computer Networks and Communication Technologies*, pp. 263–274, 2019.
[15] N.N. Misra, Y. Dixit, A. Al-Mallahi, M.S. Bhullar, R. Upadhyay and A. Martynenko, "IoT Big Data and Artificial Intelligence in Agriculture and Food Industry", *IEEE Internet Things J*, vol. 9, no. 9, pp. 1, 2020.
[16] P. S. Nishant, P. Sai Venkat, B. L. Avinash and B. Jabber, "Crop Yield Prediction based on Indian Agriculture using Machine Learning", *2020 International Conference for Emerging Technology (INCET)*, pp. 1–4, 2020.
[17] H. Nandanwar, A. Chauhan, D. Pahl, and H. Meena, "A survey of application of ML and data mining techniques for smart irrigation system," in *Proc. 2nd Int. Conf. Inventive Res. Comput. Appl. (ICIRCA)*, Coimbatore, India, Jul. 2020, pp. 205–212.
[18] R. Moeller, T. Deemyad, and A. Sebastian, "Autonomous navigation of an agricultural robot using RTK GPS and pixhawk," in *Proc. Intermountain Eng., Technol. Comput. (IETC)*, Orem, UT, USA, Oct. 2020, pp. 1–6, doi: 10.1109/IETC47856.2020.9249176
[19] Xing Yang, Lei Shu, Jianing Chen, Mohamed Amine Ferrag, Jun Wu, Edmond Nurellari, et al., "A Survey on Smart Agriculture: Development Modes Technologies and Security and Privacy Challenges," *IEEE/CAA J. Autom. Sinica*, vol. 8, no. 2, pp. 273–302, Feb. 2021.
[20] F.B. Poyen, A. Ghosh et al., "Prototype Model Design of Automatic Irrigation Controller," *IEEE Trans. Instrum. Meas*, vol. 70, pp. 9502217, 2021.
[21] P. K. R. Maddikunta et al., "Unmanned aerial vehicles in smart agriculture: Applications, requirements, and challenges," *IEEE Sensors J.*, vol. 21, no. 16, pp. 17608–17619, Aug. 2021, doi: 10.1109/JSEN.2021.3049471
[22] Merihan E. Amr, Amr A. Al-Awamry, Marwa A. Elmenyawi and Adly S. Tag Eldien, "Design and Implementation of a Low-cost IoT Node for Data Processing Case Study: Smart Agriculture," *J. Commun.*, vol. 17, no. 2, pp. 99–109, February, 2022.
[23] S. Qazi, B. A. Khawaja and Q. U. Farooq, "IoT-Equipped and AIEnabled Next Generation Smart Agriculture: A Critical Review Current Challenges and Future Trends," *IEEE Access*, vol. 10, pp. 21219–21235, 2022.
[24] P. Abdul Hafeez, G. Singh et al., "Agriculture and Healthcare: Applications and Challenges", *3rd International Conference on Smart Electronics and Communication ICOSEC 2022-Proceedings*, pp. 446–450, 2022.

Chapter 7

# Moving towards sustainable and smart agriculture

*Nidhi Sharma, Shweta Taneja and Bhawna Suri*

## 7.1 INTRODUCTION

The evolution of agriculture in India takes place over thousands of years; there is documentation of early agricultural practices in Vedas and Upanishads also. India stood at the seventh position in the world in terms of land area coverage of 328 million hectares. Out of this, 156.4 mha of land is arable, and the cultivated land area is 140 mha. As per the statistical data of 2015–16, only 48.7% of the cultivated area is under irrigation [1]. Agriculture is one of the oldest professions in India. Agriculture, along with livestock rearing, fisheries and agroforestry in total, contributes towards the nation's gross domestic product (GDP). The largest livestock herd in the world is in India, and also the biggest area covered by rice, wheat, pulses, spices, cotton and milk belongs to India. India is in second position for holding the largest cultivable land in the word and employs more than 60% of the rural population. Thus, this sector has become one of the prominent employment generators. It is the second-largest contributor of fruit and vegetables, tea and farmed fish. Even though the contribution of the agriculture sector in GDP is decreasing due to the industrial and service sectors coming into prominence, this decline is not related to agricultural productivity in any way but is considered to be a part of the structural transformation of the economy. Agriculture has undergone major changes over the years in terms of cropping patterns and crop intensity, land holding and area under cultivation. The Indian food industry is expanding with its expeditious growth of the food processing industry, making its contribution to world food trade every year. The contribution of the Indian food processing industry to the total food market is 32%, making it one of the largest industries in India. In terms of production, consumption and export, it is ranked fifth in the overall industry scenario. With the demand for processed, over-the-counter and convenience food on rise, an average urban Indian's expenses on food items is almost 50% of household expenditure on food items. With the liberalization of the policy regime, and incentives being given to the food processing sector, the investment and export potential of this factor has increased [2].

DOI: 10.1201/9781003628392-7

**Limitations faced by agriculture:**

A. Decrease in farm size of farmers: Due to a decrease in average plot size, labour productivity decreases and economies of scale cannot be achieved as mechanization is not a viable option. The average farm size of agricultural land declined from 2.28 hectares (1970–71) to 1.08 hectares (2015–16) in the last four and a half decades [1].
B. Irrigation Coverage and Water Availability: In India, a shortfall of rain in the monsoon season results in scarcity of water and subsequently followed by a low agricultural yield. Ironically, more than 60% of the country, such as Rajasthan, Andhra Pradesh, northern Karnataka, Odisha, southern and eastern Maharashtra, is drought prone. During the rain-deficit period, inhabitants resort to increased use of groundwater, which results in overexploitation and deterioration and depletion of ground water resulting in a shortfall of agricultural production. The states of Punjab, Uttar Pradesh and Uttarakhand are dependent on groundwater for fulfilling their irrigation needs. Specifically taking the case of Punjab, due to shifting in cropping patterns towards paddy, dependency on ground water for irrigation increased. The problem was further aggravated in Punjab in the year 1997 due to introduction of free electricity policy for irrigation. It is considered to be a major cause for depleting the ground water table in Punjab [2].
C. Environmental Challenges: A major challenge humanity is facing is in the form of induced climate change due to anthropogenic activities carried out across the world. Even though its origin can be traced to the natural processes, over a period of time, it intensified and became visible in the form of unnaturally extended summers and winters, frequent cyclones, and spells of drought and heavy rainfall during the off season. This natural imbalance has been backed by observed statistical data and scientific findings [3]. The observed accelerated global warming is the result of increased greenhouse gas (GHG) levels in the atmosphere. The phenomena of global warming has induced climate change, and it has disturbed precipitation, wind flow and humidity. It is also responsible for the melting of ice caps and glaciers, disturbing the biodiversity and loss of flora and fauna and the ecosystem as a whole. It has been collectively felt that decisive action at the global level should be taken to cap GHG emissions. Effect of the action taken now will be visible only after a decade [4–6].
D. Low status of Agriculture in Society: More than 70% of the world's poorest populations are agricultural labourers and make their ends meet by ploughing a fragment an of agricultural plot. The low productivity is also due to this. Most of the youths in rural areas consider agriculture to be less attractive compared to other professions. Their interest in the agricultural sector is diminished, so they increasingly migrate to cities in search of remunerative and decent employment.

E. Inadequate agricultural research and education: Education plays a key role in improving the economic status and thus contribute towards overall development. Farmers should have proper knowledge and skill sets to maximize the productivity from the sector. Due to lack of knowledge, sometimes farmers use excessive fertilizers and pesticides on the land resulting in increased salinity and toxicity of soil. Knowledge about organic farming and technology plays a key role in increasing the yield.

The main objectives of the present study are as follows:

- To know the current status of agriculture through evolution of agriculture
- To explore the role of agriculture in the Indian economy vis-à-vis food production and increasing population
- Sustainable and inclusive agriculture and challenges thereof
- Smart agriculture using AI and IoT
- To study the various types of government initiatives for the development of agricultural sector in India

## 7.2 EVOLUTION OF AGRICULTURE

Since pre-historic times, the agriculture practices followed by tribal communities were related to cultivation of land by Jhumming, burn and slash cultivation and breeding of live stocks for their survival and well-being [7]. This was a part of traditional agriculture era. This era entails the use of manpower and animals along with traditional tools like wooden ploughs and bullock carts, sickles, seeds, simple composting techniques and available water. Manual labour was involved leading to low crop yield. For a few hundred years, the Indian agricultural system has not undergone changes in terms of technological interventions [8]. During the pre-independence era, the agriculture production was mainly related to commercial crops like tea, coffee, rubber, cotton with gross neglect of food crops. With the increasing population, the demand for food crops outstripped supply, leading to food scarcity. Post-independence also, the first three 5-year plans could not bring in significant changes due to labour-intensive agriculture, with frequent spell of drought, no land reforms and no technological innovations in the picture. India faced grave famine in the 1960s. Low productivity of the agricultural sector forced India to import food grains from the United State of America and other economies [9]. Figure 7.1 shows the evolution of agriculture in India.

Green revolution in the 1960s, as shown in Figure 7.2, had brought about spectacular success as India became one of the largest producers of wheat, paddy, pulses and a variety of fruits and vegetables. It also introduced the wide use of agricultural machinery and inorganic chemicals by farmers to

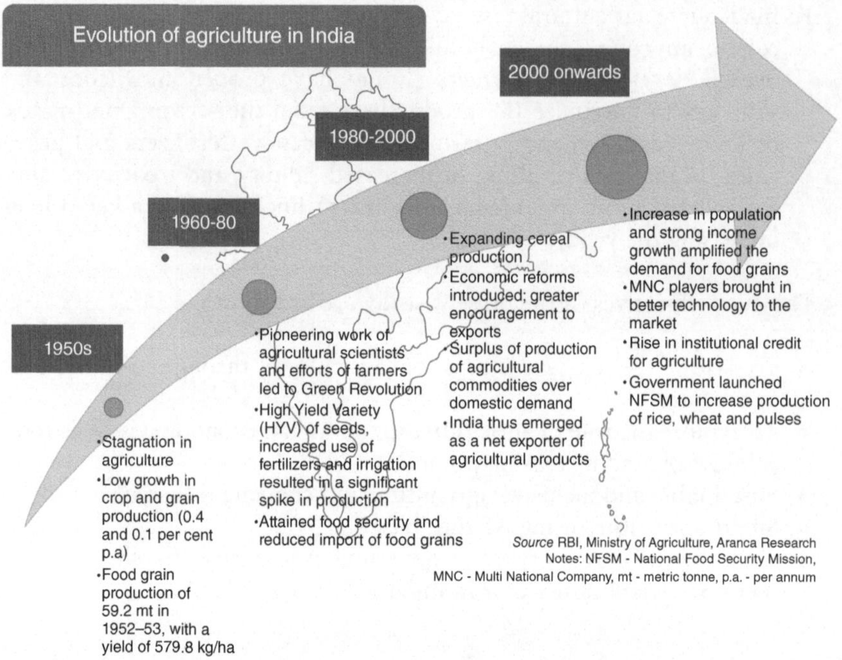


*Figure 7.1* Agricultural evolution [10].

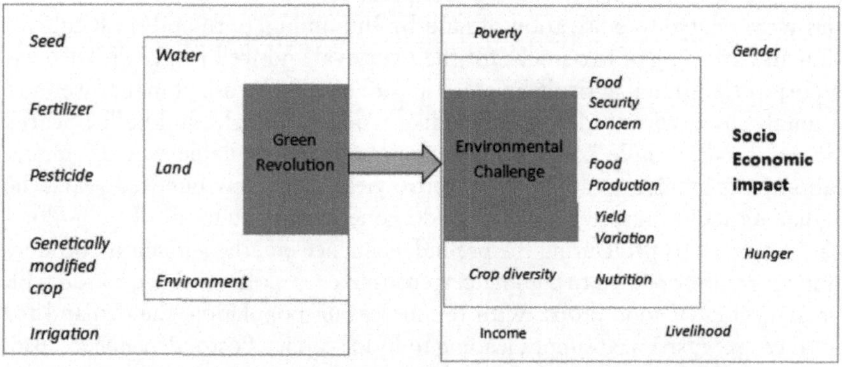


*Figure 7.2* Green Revolution 1.0 short-term boon or long-term bane.

increase the yield. The nation made great strides in agriculture with increased food production.

The green revolution focused on:

a. Increasing productivity through farm mechanization by using tractors.
b. Introduction of high-yielding hybrid varieties of seeds for increasing crop yield.

c. Using pesticides and inorganic fertilizers for improving yield and eradication of pests.
d. Construction of dams for improving irrigation. The construction of dams brought about changes in farming and improved the living condition of farmers as food crop supply was in surplus due to better irrigation facilities.

Farm Mechanization: During the 19th century, with the invention of steam engines, farm machinery came into the picture signalling the onset of Agriculture 2.0. Agricultural transformation brings about substitution of labour-intensive and livestock-based farming with machine power. Mechanized processes in combination with electrical supply have contributed tremendously with its use of only 2.6% in 1951 to about 34% in 1971 and about 88% in 2013–14. The contribution of tractors stood at 48% in 2013–14 [11]. The process of mechanization of agriculture is shown in Figure 7.3.

However, the indiscriminate use of fertilizers and pesticides and unscientific farming practices caused chemical pollution, environmental devastation and degradation of natural resources. Due to emphasis given to mono cultivation, producing subsidized and high-yielding varieties and hybrid crops, the indigenous crop was adversely impacted. One of the glaring examples is the loss of more than 1 lakh varieties of indigenous rice after the 1970s that took several thousand years to evolve [12]. This loss of species is mainly due to the focus on the production of subsidized high-yielding hybrid crops and the emphasis on monoculture by the government.

Various measures undertaken by the government to increase the production of food crops like rice, wheat, pulses led to the bumper yield, but they

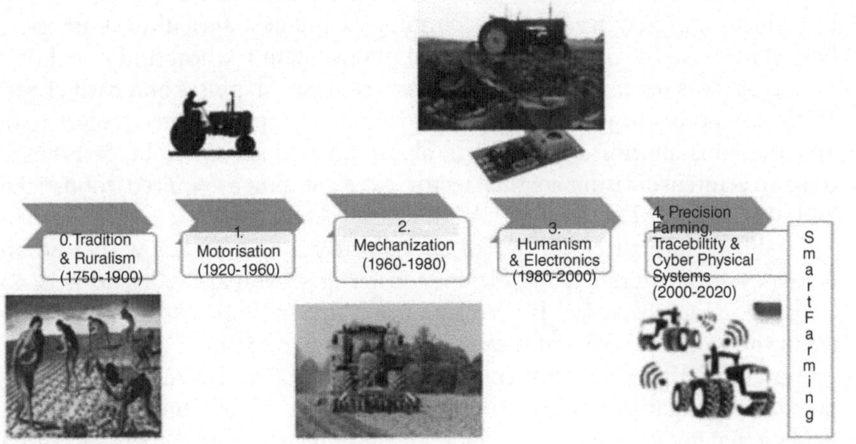


*Figure 7.3* Mechanization of agriculture.

also destroyed the diverse gene pool. The impact was faced by farmers in terms of heavy expenditure to offset these problems [13].

## 7.3 IMPORTANCE OF AGRICULTURE IN THE INDIAN ECONOMY

The agricultural sector plays a vital role in imparting food security to India. Being an agrarian society, agriculture is a key contributor to the national income and thus alleviates poverty. It plays a vital role in generating employment. Agriculture is the biggest unorganized sector of the national economy and contributes almost 90% of the unorganized workforce with more farm labourers than farmers.

In the past few decades, gradual transformation in agriculture can be observed as a result of the Indian economy being open to globalization. Globalization and liberalization led to private sector investment in the age-old agrarian sector that has conventionally been regarded as family ownership and legacy in rural India.

As a part of the government development strategy, the economic reforms introduced since 1990 have resulted in free agricultural markets, and reduction in input prices, thereby making term of trade favourable for agriculture, and external trade of agricultural goods was also liberalized in India [14].

Figures 7.4 and 7.5 show the annual growth of food grain production for pre-globalization (1971 to 1991) and the post-globalization period (1992 to 2017). In pre-liberalization time, the maximum production of 176.39 million tonnes was observed in the years 1990 and 1991 due to the green revolution and economic reforms introduced by the government. Being post-election year also, more subsidies were offered to farmers. In the post-globalization period, there was a decline in food grain production during 2002–2003; thereafter, gradual increase in production was observed. The introduction of new seeds, new technologies and new agricultural practices helped increase the production of food grains, mainly wheat and rice; however, there was no change in the production of cereal, pulses and cash crops. With easy access to water and mechanization, cropping intensity increased, and the consumption of fertilizers also increased on a per hectare basis. Resource intensification in this sector gave the much-required food, feed and fibre security [15].

As shown in Figure 7.6, Indian agriculture production has witnessed an increase and acquired self-sufficiency in terms of staple food, and it is no longer population driven. It is consumption driven. India is the fourth largest in the world in terms of total value of food production.

Agricultural production increased manifold, but it also resulted in some inadvertent negative implications viz. degradation of soil and surface water, depletion of underground water, increased salinity in irrigated areas, increased resistance to pests and weeds and greenhouse gas (GHG) emissions.

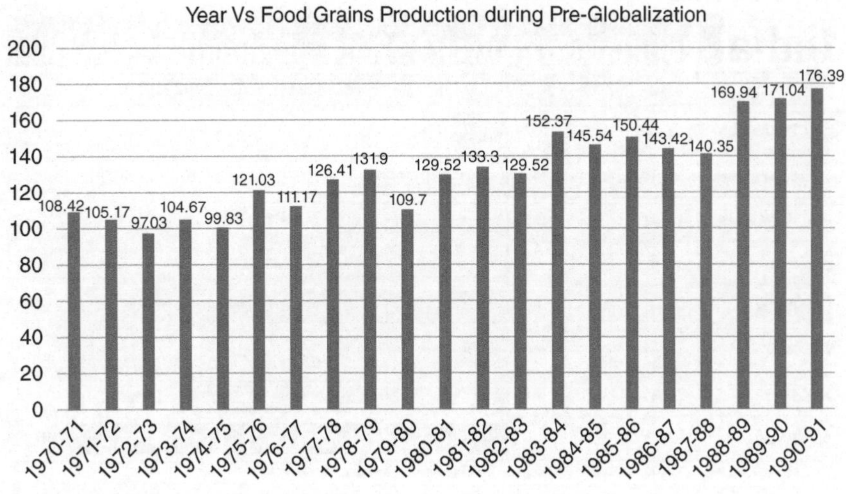


*Figure 7.4* Production of food grains in the pre-globalization period (1970–1990).

**Source: Agricultural Statistics at a Glance 2018, Government of India Ministry of Agriculture & Farmers Welfare.**

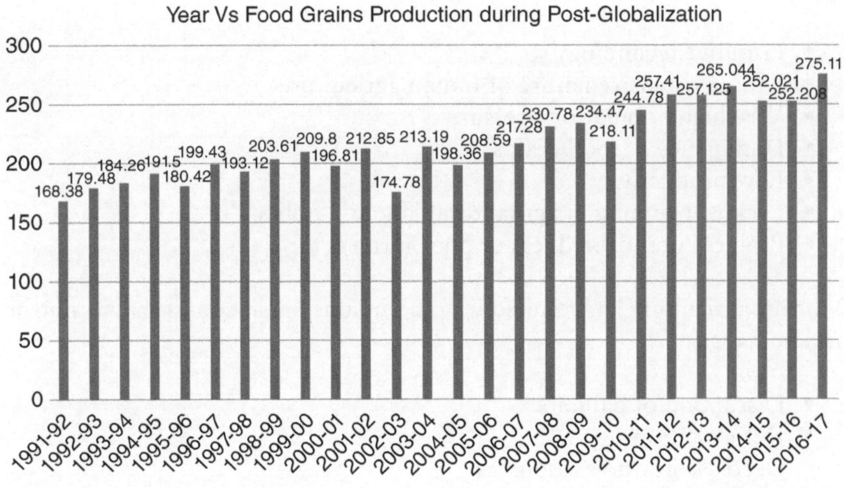


*Figure 7.5* Production of food grains post globalization (1991–2017).

**Source: Agricultural Statistics at a Glance 2018, Government of India Ministry of Agriculture & Farmers Welfare.**

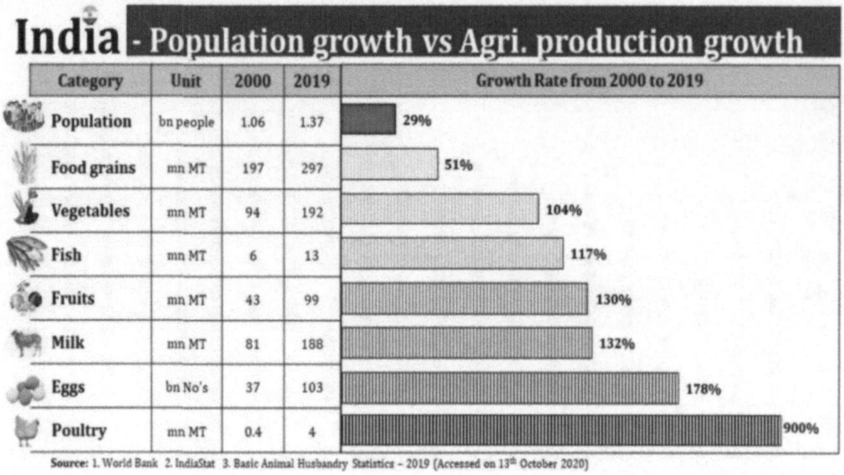


*Figure 7.6* India population growth vs. agricultural production growth.

## 7.4 NEGATIVE IMPACT OF MODERN AGRICULTURAL PRACTICES: AFTERMATH OF GREEN REVOLUTION 1.0

There are many demerits of modern agricultural practices [16]. Some of them are listed below:

- Primitive technology
- Labour-intensive nature of Indian agriculture
- Very high population pressure
- Dominance of food grains
- Low productivity
- Lack of planning in agricultural land use policy
- Problem of indebtedness of poor farmers

Modern agriculture has resulted in some serious impact on the environment and humans.

- Disruption of habitats
- Deforestation
- Decrease in water resources
- Decline in biodiversity
- Air, water and soil pollution
- Decrease in soil fertility
- Food safety issues and diminishing nutritional value of food

Therefore, there is a need for finding some sustainable and potential solutions to solve these problems. The requirement of the present time is to make agriculture sustainable which will help increase the production and profit both with the fine balance between crop, agroforestry, livestock and fisheries.

## 7.5 THE WAY AHEAD – SUSTAINABILITY IN AGRICULTURE: MOVING TOWARDS GREEN REVOLUTION 2.0

Sustainable agriculture refers to the practice of producing an adequate amount of food products, fibre and dairy/animal products by utilizing the techniques that preserve the environment, give social equity and impart food security. It aims to minimize the adverse impact or any damage to the land and surrounding ecosystem at large.

In the face of emerging environmental challenges, the agriculture sector in India needs to be climate-resilient and environmentally sustainable. Sustainability is the only option we have to improve the fertility of the soil, reducing GHG and thus mitigating climate change. By following the method of $CO_2$ sequestration, the amount of $CO_2$ in the atmosphere is decreased. Figure 7.7 shows the objectives of the sustainable and inclusive Green Revolution 2.0.

### 7.5.1 Why India needs a second green revolution

India needs the next green revolution to fulfil the growing food needs of the burgeoning billion-plus population and also to make agriculture globally

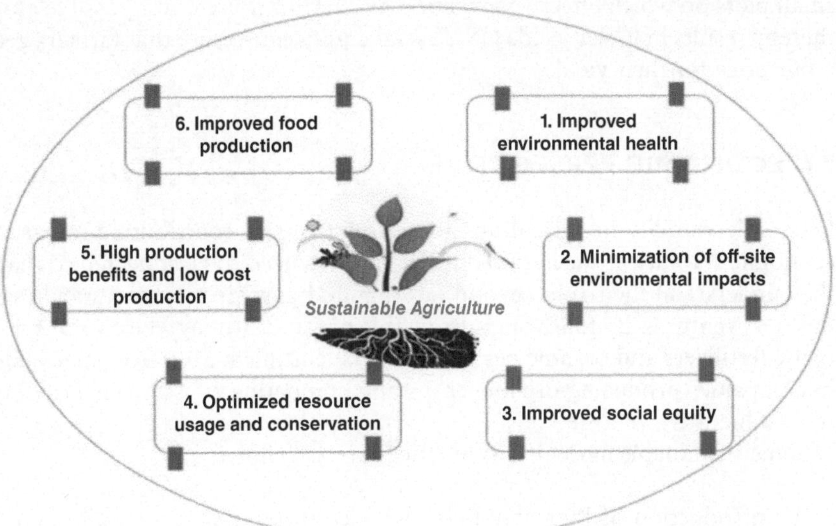


*Figure 7.7* Objectives of sustainable and inclusive Green Revolution 2.0.

competitive so that farmers can be assured of their continued livelihood and profitability. Promotion of cropping intensity and farm-specific cropping pattern and integrated farming will ensure an increase in productivity. Food production should be diversified with priority given to seed and crops which are nutrient rich [17]. Along with that, a linkage between food grain production and food processing industry is required, and it should also be linked to the market to ascertain high production benefits and good market rates.

Sustainable agriculture is significant from below-mentioned perspectives:

1. Environmental Perspective: Mitigating climate change and environmental challenges through good agricultural practices has been one of the prime objectives of sustainable agricultural practice. There is a need to adapt the alternative agricultural practices which can suit the climatic conditions of different regions. It can help in adapting to the indigenous extreme weather and help in growing resilient crops and thus help in mitigating GHG emission while maintaining or improving yields [18]. Converting fallow land to cultivable land will also help in enhancing green cover.
2. Social Perspective: Providing food security by diversifying the cultivation of food crops and cash crops. Growing high-value horticulture crops and commercial crops ensures nutritional security along with food security. Currently, more importance is given to farmers' welfare by implementing welfare schemes for improving the living condition and also to strengthen the agricultural sector.

Such welfare schemes are beneficial for farmers having fragmented land and small plots on which effective irrigation and fertilizer use is not possible and thereby results in lower yields [19]. Social equity can ensure that farmers get a fair price for their yield.

## 7.6 ECONOMIC PERSPECTIVE

Economic viability in agriculture goes a long way in contributing towards economic stability at the national level. Farmers need to be educated so that they understand their role beyond farming and participation in agriculture business ventures. By following sustainable organic farming, the cost of synthetic fertilizers and organic pesticides can be curtailed. Moreover, investing in companies producing organic crops and combating emissions is proving to be a lucrative venture.

Some sustainable agricultural practices are as follows:

1. Introduction of Biotechnology in agriculture: Biotechnology has created humongous possibilities in crop and animal production through genetic modification and hybridization of species. Crop adaptation to different climatic conditions and the production of more crops with

less land area has proved to be a boon for increasing yield. Genetic engineering has helped develop crops with very less energy requirement. Development in biotechnology will impact future agricultural technologies in terms of increasing the productive capacity with limited resources. For instance, the successful arrangement of control proteins on DNA within a cell has been reported. These advances open up a tremendous scope for nanofabrication in modern molecular biology or agricultural biotechnology of plants and animals.

2. Organic Farming: Organic farming has been used in India since ancient times, It is the method of cultivating the land using organic waste residues such as humus, food waste and agricultural residues and aquatic waste to increase the mineral content of the soil, which in turn increases the fertility of soil in the presence of worms and other microorganisms.
3. Regenerative Agriculture: Regenerative agriculture works on improving soil health and increasing crop productivity. It is an upgradation of sustainable organic farming. It aims to recharge the soil with nutrients and $CO_2$ sequestration with the help of microorganisms. It involves good agricultural practices like no-till farming, mixed cropping and crop rotation.
4. Nanotechnology and Nanobiology: Nanotechnology and nanobiology for agriculture is an emerging area that can be applied for designing and developing methods for improving the quantity as well as quality of food products. It finds applications in customizing preservation on demand, smart packaging and developing the food product which can be modified by consumers to suit their individual nutritional requirements and tastes. Thousands of nano-capsules containing flavour or colour enhancers or added nutritional elements (such as vitamins) would remain dormant in the food and would be released at will when induced by the end user. Nanoscience is exploring inexpensive nanotech applications to improve fertility and agricultural crop yield [20]. Nanoparticles may also deliver growth hormones or vaccines to livestock, or DNA for the genetic engineering of plants. A nanocomposite coating process could improve food packing by placing antimicrobial agents directly on the surface of the coated film [21].

## 7.7 BRIDGING THE GAP-SMART AGRICULTURAL PRACTICES

Following are some smart agricultural practices that are being used.

### 7.7.1 IoT-based smart farming

Smart farming uses sophisticated and modern technology and helps farmers monitor the plants remotely [22]. Smart farming is an emerging modern

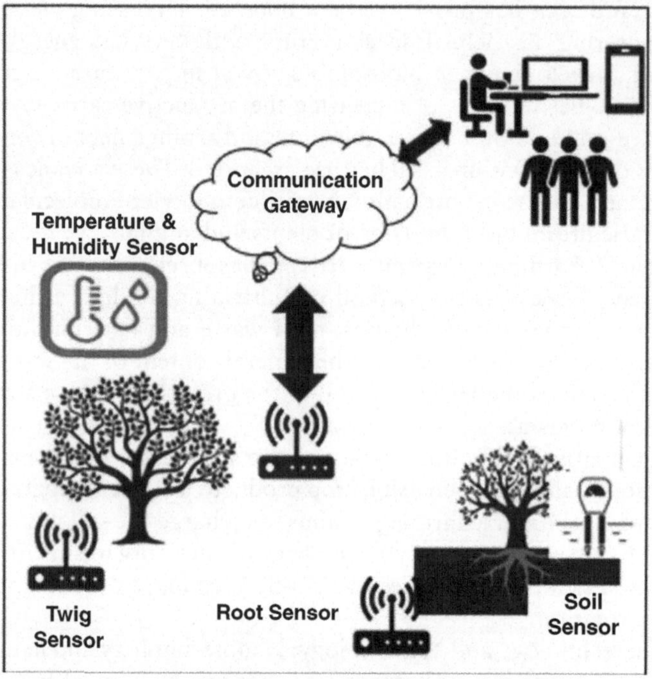


*Figure 7.8* Internet of Things (IOT)-based network for smart farming.

technology that combines information and communication technologies (ICT). It makes use of recent technologies like IoT, sensors, GPS (Global Positioning Systems), drones, robotics, data analysis tools, actuators, etc. as shown in Figure 7.8. These tools understand the need of farmers and provide the optimal solutions. These solutions help the farmer make decisions and thus improve crop productivity. Sensors continuously monitor plants and therefore can detect plant diseases or any other undesirable conditions in the early stage of plant growth.

IoT devices can collect information like climate, temperature, wind speed, soil dampness, etc. which helps farmers identify the bugs, wild plants, water level, etc. Remote sensors provide farmers with pictures and videos even remotely [23–25].

### 7.7.2 Precision agriculture

Precision agriculture is an emerging field that has automated the traditional agriculture domain. It is based on the use of IoT sensors and software that understand the needs of crops and provide them with nutrients, etc., for optimal productivity. The plant's health depends on a variety of parameters like temperature, water level, soil quality, etc. Precision agriculture particularly studies what parameters are needed, their amount and exactly where these parameters are needed for maintaining a healthy crop [26].

The IoT sensors help to collect large amounts of data related to plants' health at different growth stages. This information can be used for further decision-making. For instance, the decision to determine the amount of fertilizers needed by the plant depends on the chlorophyll amount in the plant along with the nature of soil as well as weather conditions.

The major component of precision agriculture is wireless sensor network, which is made up of multiple connected wireless nodes to gather real-time information of crops.

Every node consists of a sensor(s), a microcontroller, a radio transceiver, an antenna, along with other components that enable it to communicate with some gateway to propagate information collected by the sensor(s) [27]. IoT-based systems suffer from the limitation of increasing devices. Each node transmits data to a remote cloud, which leads to congestion in the cloud and latency with minimum power requirements, etc. A good solution is to use fog computing and edge computing methods which help reduce the traffic on the cloud.

### 7.7.3 Vertical agriculture

Vertical farming is a popular and emerging future technology that manages existing resources efficiently and produces high-quality food. It is a good method that can be used in urban areas where there is a scarcity of land and increasing pressure of growing population. The concept of vertical farming includes latest technologies like aeroponics, hydroponics and aquaponics, etc. These methods produce a wide variety of crops and thus increase production.

In vertical farming, crops are grown in old warehouses that save the need for soil as well as water. There is no constraint of any weather conditions or other natural factors [28, 29].

### 7.7.4 Smart irrigation

Water is our valuable natural resource, and it requires proper management and effective utilization. The process of agriculture involves a huge amount of water; therefore, smart irrigation methods are the need of the day. In this, low-cost sensors can be deployed in the fields to manage water for irrigation. The latest technology like IoT and WSN can be used for developing smart irrigation [30, 31].

### 7.7.5 Training and education

The process of agriculture involves nowadays the use of modern tools and techniques. Along with this, the latest information and communication techniques have also affected agriculture. Modern technologies like robots, drones, IoT devices, blockchain, etc., are being used in agriculture for real-time data collection and processing. So, farmers are required to upskill themselves or train themselves in this digital world and acquire ICT skills [32].

## 7.8 GOVERNMENT INITIATIVES

In this section, we describe the various schemes and policies that our government has implemented for promoting agriculture in the country.

There are different schemes like:

- Pradhan Mantri Kisan Samman Nidhi (PM-KISAN): Its objective is to support the Indian farmers by providing them direct income.
- Soil Health Card Scheme: It aims to improve the fertility of soil and thus enhance crop production.
- National Agricultural Market (e-NAM): It provides a uniform opportunity for farmers to do trade.
- Rashtriya Krishi Vikas Yojana (RKVY) and the Pradhan Mantri Fasal Bima Yojana (PMFBY): They aim to support farmers financially by providing them with infrastructure or raw materials and supporting them in case of crop failure, etc. [33].

Keeping the experiences of Green Revolution 1.0, the government should frame and introduce another revolution based on food security and safety, focusing on inclusive development and sustainable agriculture for future earth. Different stakeholders such as farmers, scientists, academicians, planners, policymakers and non-governmental organizations should be included in the framing of the second green revolution of the country.

Besides these schemes, photovoltaic (PV) systems were installed by the Punjab government in its subsidy scheme for the working of sprinkler and drip irrigation systems in the fields [34]. Figures 7.9, 7.10, and 7.11 show some government schemes.

The Union Minister of Home Affairs and Cooperation had launched the 'Dairy Sahakar' policy in the state of Gujarat. Another scheme named Krishi UDAN 2.0 was launched by the Ministry of Civil Aviation in 2021 to provide assistance to the farmers with the help of air transportation. This scheme was implemented at 53 airports across the country. A memorandum of understanding (MoU) was signed between the Agricultural and Processed Food Products Export Development Authority (APEDA) and ICAR-Central Citrus Research Institute (ICAR-CCRI), Nagpur, to promote exports of citrus fruits. In another initiative, 820,600 seed mini-kits were distributed free of cost in 343 districts in 15 major food-producing states to speed up the seed replacement rate and thus increase farmers' income. In September 2021, 35 crop varieties having unique traits were launched by our honourable Prime Minister Mr. Narendra Modi. Digital Agriculture

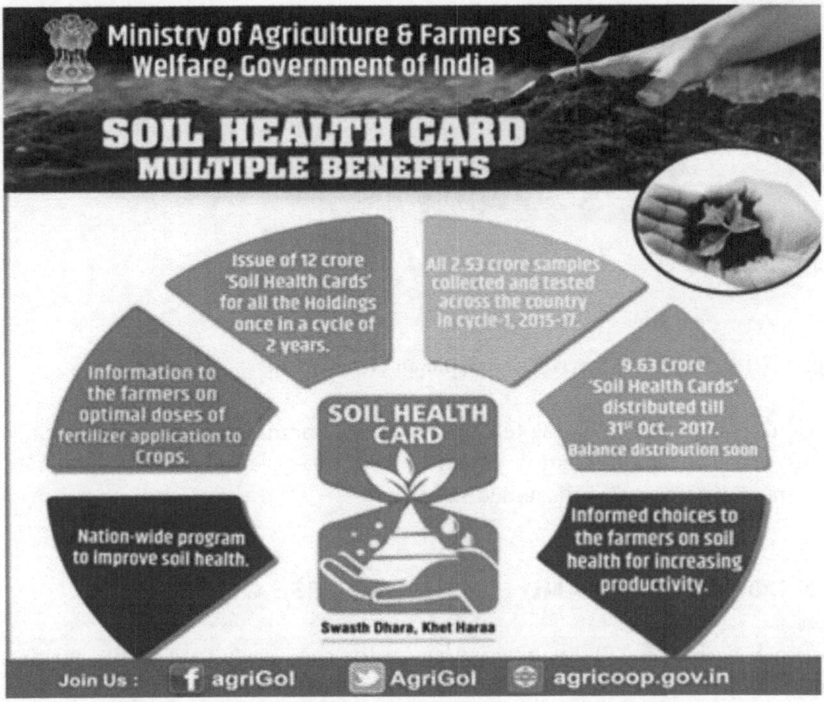


*Figure 7.9* Soil health card scheme.

*Figure 7.10* National Agricultural Market (E-NAM) scheme.

Mission has been started by the Indian government for 2021–25 for agriculture projects based on recent technologies such as drones, robots, remote sensing, AI, blockchain, etc. The Union Ministry of Agriculture and Farmers' Welfare signed five MoUs with CISCO, NCDEX e-Markets Limited, Ninjacart, Jio Platforms Limited and ITC Limited in 2021. Rs. 4,000 crore (US$ 551.08 million) was allocated towards executing Pradhan Mantri Krishi Sinchayee Yojana (PMKSY-PDMC) according to Union Budget 2021-22. An amount of Rs. 10,000 crore (US$ 1.34 billion) has been approved for a period of five years from FY21 to FY25 under Pradhan Mantri Formalization of Micro Food Processing Enterprises (PM FME). Our government has initiated the Transport and Marketing Assistance (TMA) scheme to promote

*Figure 7.11* Pradhan Mantri Kisan Samman Nidhi scheme.

agricultural exports. To make our country Atmanirbhar, different applications are built to get employment information, labour works and new schemes on farming as discussed in [35].

## 7.9 CONCLUSION AND FUTURE COURSE OF ACTION

Green Revolution 1.0 has given a lesson for the agriculture sector, farmers and policy makers that inefficient farm practices should be stopped as they lead to wastage of precious resources. For long-term sustainability in agriculture, adaptability to extreme weather conditions is required. The sector needs to be adaptable to technologies like biotechnology, nanotechnology and IoT techniques for crop protection and resilient crop production using hybrid variety of seeds. Agriculture needs to be machine and technology intensive with the deployment of sensors to monitor any anomaly in the growth pattern and convey information about any toxic constituents. Food production and processing should be interlinked, and the niche processing industry should be promoted in the global market also. In the future this sector will play a key role in fulfilling global sustainable development goals of the United Nation specifically related to SDG-1,2,3,8,10,12,13 and 15. Keeping that in mind, the policy makers and government should strive towards the next green revolution focused on inclusive development and food security with the involvement of scientists, biotechnologists, farmers, environmentalists and academicians.

## REFERENCES

[1] Press Information Bureau, Government of India, Ministry of Agriculture & Farmers Welfare, 03-March-2020 20:24 IST.

[2] Jain, Rajni, Kishore, Prabhat, & Singh, Dhirendra Kumar, Irrigation in India: Status, challenges and option, *Journal of Soil and Water Conservation* 18(4): 0–0, October-December 2019

[3] IPCC. *Climate Change 2013: The Physical Science Basis Working Group I Contribution to the IPCC 5th Assessment Report Intergovernmental Panel on Climate Change*; IPCC: Geneva, Switzerland, 2013].

[4] European Commission. Communication from the Commission to the European Parliament, The Council, The European Economic and Social Committee and The Committee of the Regions. *Forging a Climate-Resilient Europe—The New EU Strategy on Adaptation to Climate Change; European Commission*: Brussels, Belgium, 2021.

[5] Houghton, J. *Global Warming*, 3rd ed.; Cambridge University Press: Cambridge, UK, 2004. 10.

[6] Cowie, J. Climate Change, *Biological and Human Aspects*; Cambridge University Press: Cambridge, UK, 2007.

[7] Tekinerdogan, B. *Strategies for Technological Innovation in Agriculture 4.0. Reports*; Wageningen University: Wageningen, TheNetherlands, 2018.

[8] Lockwood, B., Mukherjee, P.K., & Shand, R.T. *The High Yielding Varieties Programme in India, Part 1. Programme Evaluation Organization, Planning Commission of India and Department of Economics, and Research School of Pacific Studies*, Australian National University, Canberra. 1971.

[9] Source: Press Information bureau Ministry of Agriculture & Farmers Welfare, Achieving Aatmanirbharta in Agriculture Posted On: 11 NOV 2022.

[10] Kumari, P. L., Reddy, G. K., & Veeraiah, A. (2019). Challenges and opportunities in marketing farm produce: A review. *Journal of Pharmacognosy and Phytochemistry*, *8*(1), 1022–1029.

[11] Chand, R. et al. (eds.) (2022), Indian Agriculture Towards 2030, *India Studies in Business and Economics*, https://doi.org/10.1007/978-981-19-0763

[12] The Hindu; 2012. http://www.thehindu.com/news/national/karnataka/from-110000-varieties-of-rice-to-only-6000-now/article3284453

[13] Eliazer Nelson, A.R.L., Ravichandran, K. & Antony, U. The impact of the Green Revolution on indigenous crops of India. *J. Ethn. Food* 6, 8 (2019). https://doi.org/10.1186/s42779-019-0011-9

[14] Chitra, M., & Malarvizhi, Chinnasamy Agamudai Nambi. "An Empirical Analysis on India's Food Grain Cultivation, Production and Yield in Pre & Post Globalisation." *Shanlax International Journal of Economics*, vol. 8, no. 3, 2020, pp. 1–10.

[15] https://economictimes.indiatimes.com//news/economy/agriculture/share-of-agriculture-in-indias-gdp-declined-to-15-in-fy23

[16] Rohila, A. K., Maan, D., Kumar, A., & Kumar, K. (2017). Impact of agricultural practices on environment, *Asian J. Microbiol. Env. Sc*, *19*(2), 145–148.

[17] Swaminathan, M. S., 2000, An Evergreen Revolution, *Biologist*. 47, pp. 85–89.

[18] Prandecki, K., Wrzaszcz, W., Zieliński, M. Environmental and Climate Challenges to Agriculture in Poland in the Context of Objectives Adopted in the European Green Deal Strategy. *Sustainability* 2021, 13, 1031.

[19] Yadav, Shikha, & Anand, Subhash, Green Revolution and Food Security in India: *A Review,NGJI, An International Peer Reviewed Journal NGSI-BHU*, ISSN: 0027-9374/2019/1715 Vol. 65, No. 3, September, 2019.

[20] Opara Linus, U. (2004), "Emerging Technological Innivation Triad for smart Agriculture in the 21st Century. Part 1 Prospects and Impacts of Nanotechnology in Agriculture", *Agricultural Engineering International: the CIGR Journal of Scientific Research & Development*, Invited Overview http://www.cigrjournal.org/index.php/Ejounral/article/view/554/548 down loaded on 11-05-2011].

(http://knowledge.cta.int/en/layout/set/iprint/Dossiers/S-TIssues-in-Perspectie/Nanotechnology
[21] DR. SEEMA SINGH,Achieving Second Green Revolution through Nanotechnology in India, https://eands.dacnet.nic.in/Publication12-12-2012/1485-jan12/1485-1.pdf
[22] O'Grady, M.J.; O'Hare, G.M.P.. Modelling the smart farm. *Inf. Process. Agric.* 2017, 4, 179–187)
[23] Naresh, M., & Munaswamy, P. (2019). Smart agriculture system using IoT technology. *International journal of recent technology and engineering*, *7*(5), 98–102.
[24] Sushanth, G., & Sujatha, S. (2018, March). IOT based smart agriculture system. In *2018 International Conference on Wireless Communications, Signal Processing and Networking (WiSPNET)* (pp. 1–4). IEEE.
[25] Raj Kumar, G., Chandra Shekhar, Y., Shweta, V., & Ritesh, R. Smart agriculture—Urgent need of the day in developing countries. *Sustain. Comput. Inform. Syst.* 2021, 30, 100512).
[26] Shafi, U., Mumtaz, R., García-Nieto, J., Hassan, S. A., Zaidi, S. A. R., & Iqbal, N. (2019). Precision agriculture techniques and practices: From considerations to applications. *Sensors*, *19*(17), 3796.
[27] Wang, N., Zhang, N., & Wang, M. Wireless sensors in agriculture and food industry—Recent development and future perspective. *Comput. Electron. Agric.* 2006, *50*, 1–14.
[28] Al-Kodmany, K. *The Vertical City: A Sustainable Development Model*; WIT Press: Southampton, UK, 2018.
[29] Benke, K., & Tomkins, B. Future food-production systems: Vertical farming and controlled-environment agriculture. *Sustainability: Science, Practical and Policy*. 2017;13:13–26.
[30] García, L., Parra, L., Jimenez, J. M., Lloret, J., & Lorenz, P. (2020). IoT-based smart irrigation systems: An overview on the recent trends on sensors and IoT systems for irrigation in precision agriculture. *Sensors*, *20*(4), 1042.
[31] Nawandar, N. K., & Satpute, V. R. (2019). IoT based low cost and intelligent module for smart irrigation system. *Computers and electronics in agriculture*, *162*, 979–990.
[32] Pogorelskaia, I., & Várallyai, L. (2020). Agriculture 4.0 and the role of education. *Journal of Agricultural Informatics/Agrárinformatika Folyóirat*, *11*(1).
[33] Tripathi, G., Dhodia, A., Giri, A., Rathore, V., Verma, A., Shukla, A., & Verma, L. K. (2023). Government Agriculture Schemes in India: A Review. *Asian Journal of Agricultural Extension, Economics & Sociology*, *41*(11), 58–67.
[34] Raza, F., Tamoor, M., & Miran, S. (2021). Socioeconomic and climatic impacts of photovoltaic systems operating high-efficiency irrigation systems: A case study of the government subsidy scheme for climate-smart agriculture in Punjab, Pakistan. *Engineering Proceedings*, *12*(1), 36.
[35] Suri, B., Taneja, S., Dhingra, G., Goyal, A., Sharma, B. (2023). Atamnirbhar Gaon—An Inhouse Employment Tool for Migrant Workers. In: Gupta, D., Khanna, A., Hassanien, A.E., Anand, S., Jaiswal, A. (eds) *International Conference on Innovative Computing and Communications. Lecture Notes in Networks and Systems*, vol 492. Springer, Singapore. https://doi.org/10.1007/978-981-19-3679-1_29

Chapter 8

# Integrating Green IoT with Precision Agriculture

## A Paradigm Shift in Sustainable Farming Practices

*Diksha Jain, Jolly Parikh, Rachna Jain and Vassilis C. Gerogiannis*

## 8.1 INTRODUCTION: BACKGROUND AND DRIVING FORCES

Conventional greenhouse management techniques rely heavily on manual observation and intervention by growers or agricultural workers. This manual approach [1] involves regularly keeping an eye on the greenhouse's numerous environmental conditions, including temperature, humidity, light intensity, and $CO_2$ levels, and making adjustments as needed to create optimal growing conditions for plants. However, this process is inherently time-consuming, labor-intensive, and prone to human error.

First of all, in order to evaluate environmental conditions by manual observation, gardeners must physically examine various sections of the greenhouse. This may require a long time, particularly in big greenhouse operations with many plants grown across wide growth areas. Additionally, the burden of growers may increase due to variations in monitoring frequency based on plant type, development stage, and time of day.

Second, sustaining optimal growth conditions may not always be possible with actions based only on manual monitoring [2]. Because human judgment is subjective, it can be impacted by a variety of factors, including experience, weariness, and how each person interprets surrounding cues. Because of this, farmers could unintentionally miss minute alterations in the environment or neglect to react quickly enough to changes in it, which might result in less-than-ideal growth conditions and decreased plant yield.

Moreover, there are a lot of intricate relationships and complicated interactions between the many climatic components in a greenhouse. For instance, plant transpiration rates, water intake, and susceptibility to pests and diseases can all be impacted by the interaction between temperature and humidity levels. Similarly, Vialet-Chabrand et al. [3] have shown that photosynthesis, carbon uptake, and total plant growth depend heavily on light intensity and $CO_2$ concentrations. It may be difficult to attain this balance only by hand. Rather, careful management and ongoing monitoring are needed to generate the ideal microclimate for plant development.

DOI: 10.1201/9781003628392-8

The work focuses on developing an Edge AI program to improve greenhouse plant growth. The main goal of this research is to give the system the ability to evaluate the greenhouse's environmental conditions to determine their suitability for plant development. The research explores the link between technology and the complex requirements of modern agriculture by incorporating artificial intelligence (AI) capabilities into an embedded system.

An HTU21D humidity and temperature sensor is integrated into the system's core. This sensor plays a key role in data collection, acquiring important environmental data continually with a focus on humidity and temperature readings. These two factors have a direct and significant impact on the health of plants by affecting a number of physiological and metabolic processes. The Edge AI gadget can decide whether the greenhouse atmosphere is suitable for optimum plant growth by utilizing data from this sensor.

The prototype integrates a Waveshare 1.3-inch OLED screen into the design to improve user engagement and offer informative feedback. This display element offers real-time visualizations of gathered data and assessments of greenhouse conditions, acting as a window into the system's decision-making processes. The gap between practical greenhouse management and the intricate technicalities of AI algorithms is filled by this interface.

The STM32 NucleoF401RE microcontroller, which serves as the system's central processing unit, facilitates this intelligent environment. This microprocessor acts as the cognitive core of the system, coordinating the HTU21D sensor, Edge AI algorithms, and the OLED display. The system's computational capabilities allow it to process AI models, process incoming data, and produce real-time reactions based on evaluated conditions.

The team's work is centered on the incorporation of AI algorithms, microcontroller technology, and sensors to improve greenhouse plant production. The intention is to improve agricultural technology through the use of more practical and effective methods.

## 8.2 LITERATURE REVIEW

This section discusses earlier studies that have been conducted in the same area and use approaches and applications that are like the team's work. The team's goal in reviewing these past publications is to situate their findings within the larger framework of relevant investigations. While the approach is like previous work, the team's contributions are in streamlining and simplifying the process, hence improving its practicability.

Three main subsystems make up the work in ref. [4]: ambient condition monitoring and warning, illness prediction, and automated watering. They monitor environmental factors via a camera, temperature-humidity sensors, and soil moisture sensors to obtain critical data. The gadget initiates automated plant watering and promptly warns users of inadequate temperature and humidity levels when soil moisture levels are low. They include machine learning (ML) techniques in their analysis of leaf photo data for illness diagnostics using the Raspberry Pi and Internet of Things technologies. The team's concept is supported by this solution, which makes use of temperature-humidity sensors to assess the surroundings and display the findings on an LED screen.

In order to achieve this, research conducted by Christian von Zabeltitz in *Integrated Greenhouse Systems for Mild Climates: Climate Conditions, Design, Construction, Maintenance, Climate Control* [5], as referenced in the study by C. J. Brien et al. [2], discusses suitable greenhouse structures and their designs, tailored to meet the infrastructure requirements and local temperatures in subtropical, tropical, and dry regions. As optimal conditions for agricultural growth have become increasingly important, these designs focus on enhancing environmental factors such as temperature, humidity, and light to ensure better crop yield and sustainability.

In [3], the researchers have developed a system that integrates sensor elements into a Raspberry Pi to monitor changes in temperature, humidity, and soil moisture. The system may then notify the user using a mobile application. Data on agriculture may be analyzed using sensor values that are entered into a Raspberry Pi MySQL database. The built authorized mobile application may be used to track parameters and obtain the most recent news and schemes related to agriculture, market crop rates, and meteorological data, among other things. The government market agricultural prices offered will shield farmers from the avaricious intermediaries.

The goal of the experimental design for the system that the authors have suggested [6] is to increase agricultural yield in a controlled setting and increase crop yield in a regulated setting. Sensors for light, temperature, humidity, and soil moisture are employed to gather important data for this purpose. Decisions about needs are made using an Arduino Mega 2560 and threshold values for certain environmental factors. Then, based on the environmental conditions, the control system creates control signals to activate the sprinklers and exhaust fans. With an Android application, you may monitor crop health and environmental conditions. Furthermore, if necessary, actuators may also be operated via the Android application. Red and blue light emitting diodes (LEDs) are employed in a unique combination to improve crop growth and preserve crop health. The greenhouse system is self-sufficient since it is powered by solar panels.

A greenhouse is designed specifically to accommodate the kind of plant it wants to grow [7]. Thus, the structure differs based on the kind of plant and its size. Despite providing the ideal habitat for plants, it requires human maintenance to maintain the house's ideal conditions, such as ventilation. People may cultivate plants more easily with the help of an automated greenhouse. When they're not home, it helps to keep an eye on things. This paper's primary goal is to automate the greenhouse and keep an eye on the condition of the indoor environment in order to reduce the amount of human care required for the plant. The prototype greenhouse structure has been built, with one unit integrated with the sensors. Adriano Uno microcontrollers are used in the construction of the control system. When rain is detected, servo motors are utilized to push the roof. Additionally, a 12-volt fan is added, and it activates when the temperature rises too high. Five distinct locations are used to imitate the produced prototype, and the outcomes are examined.

Similarly, authors of [8] mention uses of Raspberry Pi and Arduino to monitor, regulate, and coordinate several parameters that affect agricultural yield, including humidity, $CO_2$ levels, light intensity, soil moisture, and temperature. Real-time data gathering from the Smart Greenhouse and display on the ThingSpeak platform are made possible by the Internet of Things. This article suggests creating a highly sophisticated and varied version of the already popular models by integrating hydroponics and vertical farming into a fully automated greenhouse with superior security features and surveillance.

It also becomes useful when authors in [9] track greenhouse conditions using temperature-humidity sensors and soil moisture sensors.

The use of an optimal scheduler for ongoing control and observation of every stage of the crop cycle is investigated in the study [10]. Agronomists have been adopting manual or semi-auto data collection methods in recent years to monitor several aspects of the greenhouse environment, including temperature, humidity, carbon dioxide ($CO_2$), and soil moisture [11]. Bluetooth-enabled alarm systems are used for greenhouse temperature warnings.

### 8.2.1 Smart agriculture using IoT applications

The study suggested a greenhouse agriculture control system and remote sensing of agricultural characteristics. The objective is to regulate $CO_2$, soil moisture, temperature, and light. The greenhouse windows and doors are controlled based on crop conditions once a quarter of the year, depending on the soil moisture levels. Increasing the produce and offering organic farming are the goals. The system demonstrates the ability to remotely manage $CO_2$, soil moisture, temperature, and light levels [12]. For greenhouses, where temperature and humidity are crucial for plant growth, [13] also measures stress factors to predict plant development.

The goal of this new research is to turn agricultural greenhouses into smart greenhouses by deploying them on a modest scale [14]. They will assist in keeping an eye on the conditions of the greenhouse, managing irrigation and watering, taking pictures with installed cameras, and forecasting plant illnesses based on leaf statistics gathered. This study focuses on the development of a suggested system architecture and design for an appropriate IoT-based monitoring system for environmental conditions, water irrigation system management, and an efficient way to identify leaf diseases in plants grown in greenhouse environments.

A method for changing and reorienting agricultural systems to promote food security in light of the new climate change realities is called "climate-smart agriculture," or CSA [15]. The majority of the world's poor are among those whose livelihoods depend on agriculture, and their susceptibility is increased by the widespread changes in rainfall and temperature patterns that pose a danger to agricultural productivity. Climate change upends food markets, putting the food supply in danger for the whole population. Enhancing farmers' ability to adapt, together with boosting agricultural production systems' resilience and resource efficiency, can lessen threats.

Furthermore, it is simple to incorporate this novel technology into unmanned aerial vehicles (UAVs) that are sold commercially. Two distinct measuring methods have been created in order to properly categorize the ploughing techniques [16]. Experiments verify that the suggested approach can accurately classify the ploughing depths in the field.

Additionally, the majority of applications for continuous monitoring are moving from wired networks to wireless ones [17]: low-power, short-range communications [18]. Maintaining constant observation and control over the climate is essential in a greenhouse. The design and performance study of a wireless sensor module for flexible greenhouse monitoring is presented in this work [19]. The module is based on low-power ZigBee communication modules and general-purpose microcontrollers.

Similarly, it helps incorporate greenhouses into climate-sensitive parameters, fostering a microclimate that supports plant growth while conserving water and energy. A sophisticated fuzzy logic-based control system [20] was implemented and improved by optimizing the relationship between humidity and temperature. In another approach, [21] uses sensors such as the DHT11 and soil moisture sensor, along with a microcontroller, Xbee, and the ThingTalk platform for data storage and retrieval, to transmit temperature, humidity, and soil moisture data to the cloud. A real-world application of this wireless sensor network was demonstrated in the Malaysian jungle [22].

Authors of this research mention the use of cloud computing, AI, and a wireless sensor network [23], further enhancing the study of IoT-enabled technologies as discussed in [24], which help in monitoring and managing greenhouse conditions and plants, as well as forecasting tomato growth rates. This approach contrasts with ours [25], where Aquino, Ballado, and Bautista focus on implementing a wireless sensor network [26] with multiple Arduino-based farming multi-sensor tools to monitor a small farm. Our approach, in contrast, utilizes the ESP-WIFI-MESH protocol, arranging sensor nodes for small-scale farm-area monitoring in an autonomous, scalable, self-forming, and self-healing mesh network architecture. Each sensor node is equipped with several sensors to collect data. The numerous sensors attached to a sensor node will gather the plant growth characteristics, which include soil moisture, temperature, relative humidity, and ambient sunshine intensity, as input for this study. A compilation of historical data that may be viewed and examined by farmers or other pertinent organizations is the output.

As technology advances, ML becomes more and more important in order to facilitate intelligent growth. Three distinct apple diseases—Alternaria, apple black spot, and apple leaf miner pest—that manifest on leaves were chosen for this study's image processing approach detection. In this study, three soft computing approaches—support vector machines (SVMs), artificial neural networks (ANNs), and illness classification—are presented. After being sampled, the diseased leaves were sent to a lab where controlled lighting was used to take pictures of the leaves [27]. K-means clustering was then used to identify affected areas. After processing, characteristics were retrieved from the photos. For the categorization of diseases, the SVM method outperformed the ANNs.

Similarly, the intelligent monitoring system may prevent disease and insect pests, lower the need for pesticides and fertilizers, produce high-quality food, and anticipate the harsh greenhouse conditions. Greenhouse monitoring systems are gaining more and more attention from researchers, but putting them into production efficiently and appropriately is a significant problem. This paper aims to perform a systematic review of the intelligent greenhouse monitoring system; support the data transmission and server processing subsystems by defining, classifying, and describing the greenhouse's environmental parameters; and investigate the overall architecture of the system [28]. The paper conducts a comparative analysis based on the attributes of each system component to determine its development trend. It also summarizes the popular technologies of today and works on how the intelligent monitoring system will develop in the future. Finally, it offers support for the ongoing research on greenhouse monitoring systems. It was discovered that multiparameter monitoring helps to produce efficient greenhouse control, and for data transfer in the environment both inside and outside the greenhouse, wireless technology has increasingly replaced the cable mode.

Specifically, deep learning, big data, and other cutting-edge technologies utilized in greenhouse monitoring are regarded as significant advancements that enhance autonomous greenhouse management and optimize the energy efficiency of greenhouse building.

Climate change helps with solutions to lessen uncertainty since, in many locations, it creates significant uncertainty about future water supply and agricultural sustainability [29, 30]. This strategy is similar to [31] and emphasizes the importance of these elements in contemporary greenhouse management.

Using the same concepts, [32] have presented an aquaponics system that works with Android and IOT to maintain the plant growth. In [33] in the event of any unusual circumstance, information will be forwarded to the gardener or agriculturist via an Android app, and a suitable remedy will be supplied.

This paper introduces a simple and basic prototype for smart greenhouse plant growth detection systems. The focus is on offering a practical and useful method for observing and enhancing plant growth in greenhouse settings. This prototype simplifies the strategy while guaranteeing its viability and effectiveness, making it available to a wider spectrum of greenhouse producers and enthusiasts. It intends to promote the wider adoption of smart greenhouse practices by streamlining the technology.

## 8.3 OVERVIEW OF SYSTEM HARDWARE

The system uses three primary parts, the microcontroller, temperature-humidity sensor, and the LED screen. The functions of the above three are explained in this section.

### 8.3.1 Microcontroller

The team has built the entire work around the STM32 Nucleo F401RE microcontroller. This is the main brain of the work that collects data, analyses it, and sends the result to the output screen. It has an ARM cortex-M4 processor. It has a maximum frequency of 84 MHz. It has 512 kB of Flash memory and 96 kB of SRAM [31].

### 8.3.2 Temperature-humidity sensor

The team has used the HTU21D temperature and humidity sensor to collect the data of temperature and humidity from the surroundings and send it to the STM microcontroller. It works on the I2C framework. It works on 3.3 V. Its operating range is –40 °C to 125 °C and 0% to 100% RH [32]

### 8.3.3 LED screen

The team has used Waveshare's 1.3 inch OLED screen sh1106. It works on the SPI framework [33]. It is used here to display the values of temperature and humidity and tells if the environment is optimal or not.

Figure 8.1 shows how the data is transmitted across the system. First the sensor sends the data to the microcontroller. Microcontroller analyses it and sends the output to the LED screen.

### 8.3.4 Interfacing microcontroller STM32 and HTU21D sensor

The HTU21D is a humidity and temperature sensor that has an operating range of –s40°C to 125°C and 0% to 100% RH at 3.3V supply as shown in Figure 8.2.

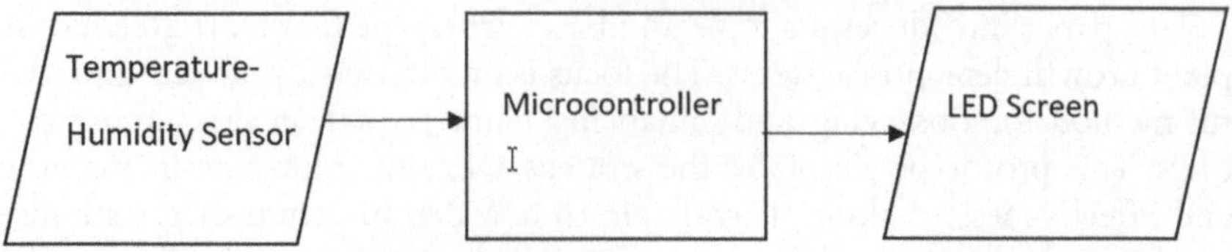


*Figure 8.1* Flowchart explaining flow of data.

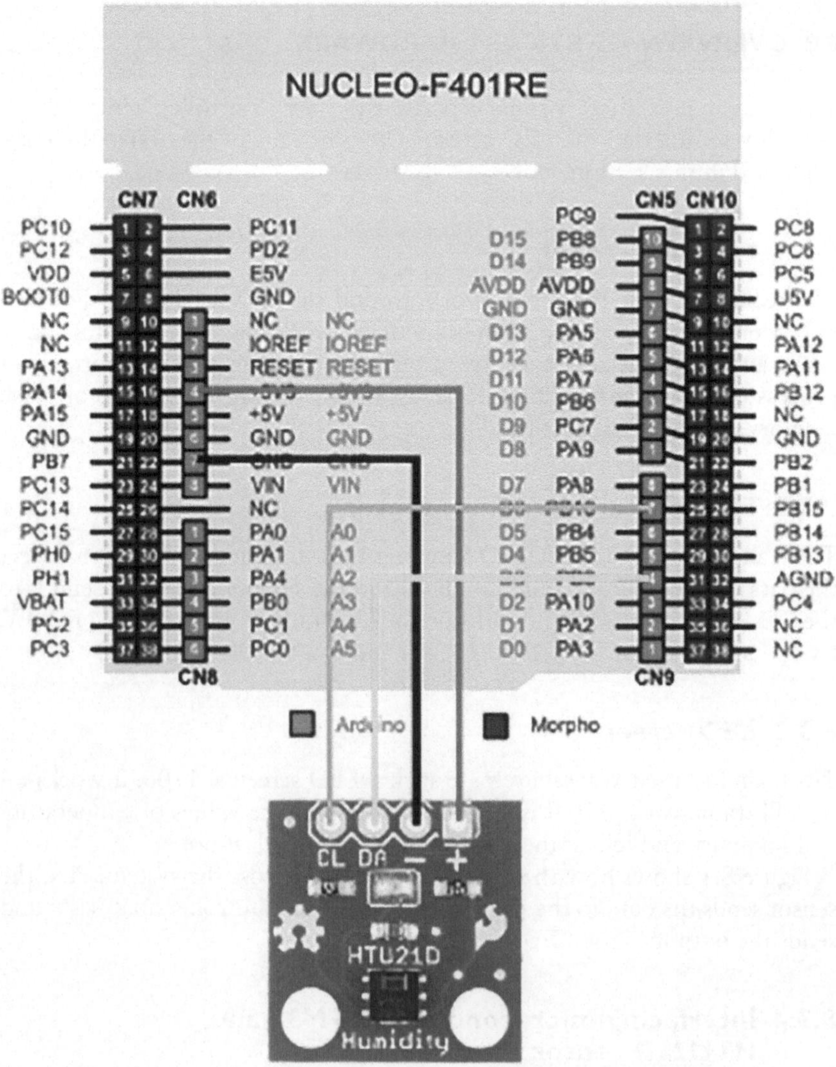


*Figure 8.2* Connecting microcontroller STM32 and HTU21D sensor.

- It works on the I2C protocol. Hence only two connections SDA (serial data) and SCL (serial clock) are needed.
- It works on the fast mode in I2C with a speed of 400 kb/s.
- The SCL line must be attached to a pull-up resistor internally or externally.

## 8.4 SOFTWARE DESIGN

As this work was built around the STM32 Nucleo F401RE microcontroller [34], it was entirely developed using the STM32 Cube IDE, and the ML model used for checking optimality was developed using NanoEdge AI Studio.

Figure 8.3 shows how the program works. First the data is collected in the form of buffers. These buffers are arrays of data in an organized manner separated by commas or spaces. The data here is a dual axis or of two forms, that is, temperature and humidity [35]. The standard size of a buffer is 256 data points. So here with two axes, each buffer has 512 values. Now this buffer is sent to the ML model deployed, which analyzes the buffer to give the output. Now this output is sent to the LED screen [36].

The raw data of temperature and humidity is collected through the I2C protocol. Now this data is converted to usable values of temperature in °C and humidity in % relative humidity through the following formula.

$$\text{Temp} = -46.85 + 175.72 \times \left(\text{Raw value} / 2^{16}\right) \tag{8.1}$$

$$\%\text{RH} = -6 + 125 \times \left(\text{Raw value} / 2^{16}\right) \tag{8.2}$$

Then this data is added to the buffer, and the process is repeated till the buffer is full. Figure 8.4 explains how a buffer is created.

The ML model used here was found using NanoEdge AI Studio software. This software analyzes the datasets given to it and finds a reliable ML model for the specified application. The team used an n-class classification application here. A total of 65 datasets were used out of which 5 were the real-life datasets taken for non-optimal conditions and 60 were artificially created using Python as the range of optimality was too broad to be taken in a few datasets.

When these datasets are added to the software, it provides a graphical analysis of the data and shows how the machine learning (ML) model fits the data into different classes. The software then identifies a list of ML models suitable for the given datasets, including details such as the model type, accuracy, and the memory and flash size required for deployment. Additionally, it generates a confusion matrix that summarizes the model's performance based on the datasets.

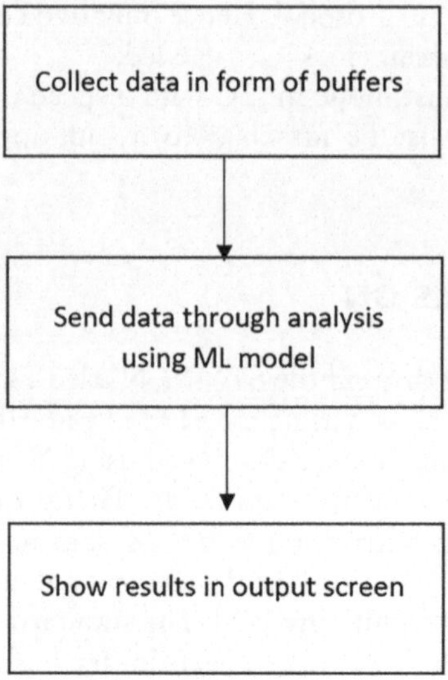


*Figure 8.3* Flowchart explaining the code flow.

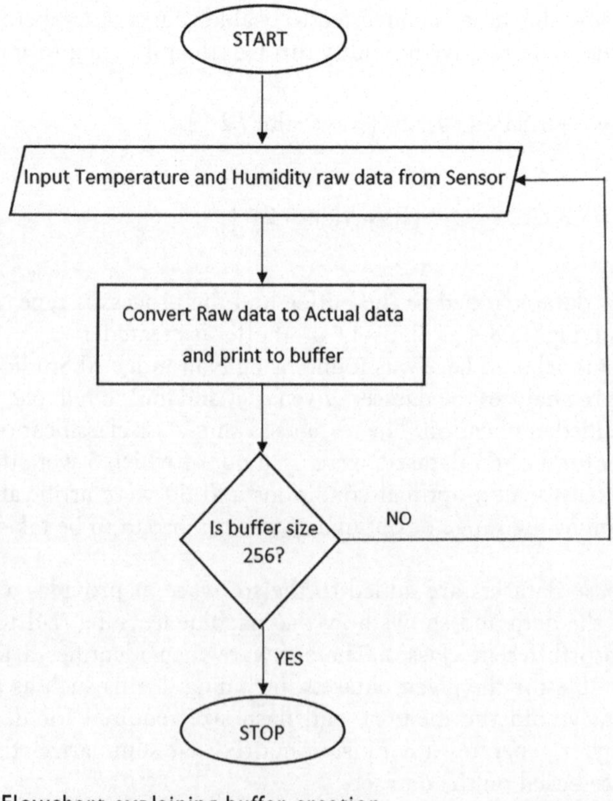


*Figure 8.4* Flowchart explaining buffer creation.

The software found 54 suitable models in a benchmarking process of two hours out of which the model found most suitable for the given datasets was a random forest model. It is an ML model that combines multiple decision trees to make more accurate predictions.

Figure 8.5 shows how much data is correctly predicted into the respective classes. The data points between 0% and 100% that deviate significantly are considered outliers that show the incorrectly predicted data.

As shown in Figure 8.6, the confusion matrix tells us how much data the model could fit in the correct class. For example, if the value is 19 that means 19 values were predicted correctly in the true class. Here the labels are from A to Z to 10 to 19 to 1A to 1Z to 20 to 23, out of which the last five classes are the non-optimal conditions. As seen in Figure 8.6, while some values are inaccurately predicted, this does not affect the outcome of the study as all these classes are considered the same class that is non-optimal environment.

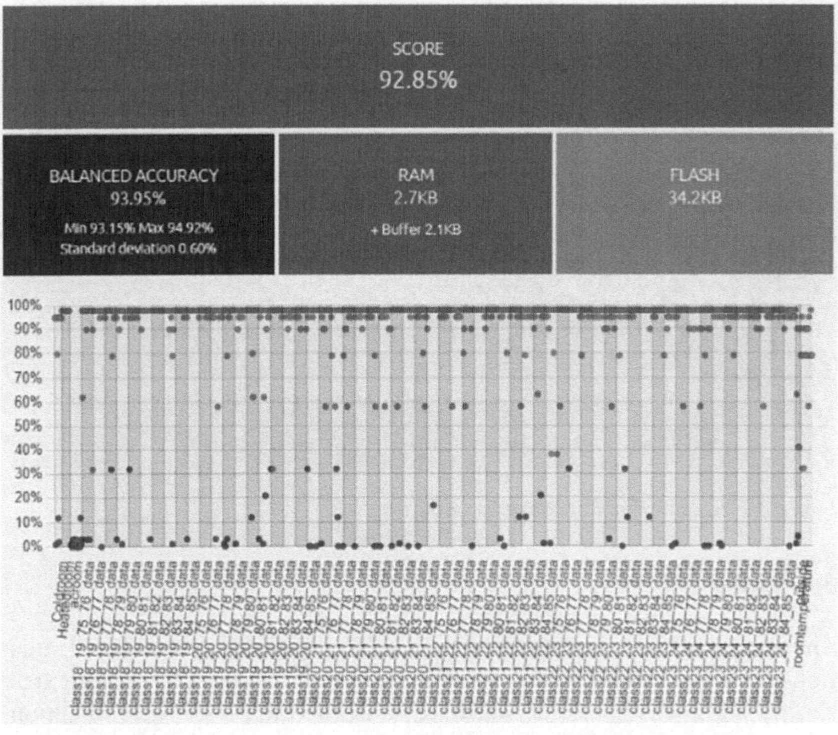


*Figure 8.5* Information about the selected machine learning model.

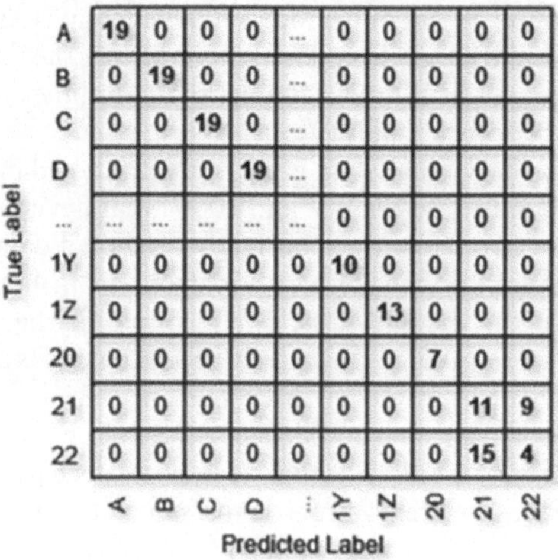


*Figure 8.6* Confusion matrix.

*Figure 8.7* Results for non-optimal condition.

## 8.5 RESULTS

The system was put through extensive testing methods that included situations for both optimal and non-optimal datasets. The results of this extensive testing were remarkably accurate in identifying the climatic conditions necessary for greenhouse plant development. Figure 8.7 illustrates a scene in which the system's LED screen shows a snapshot of non-optimal circumstances. The system's ability to identify and communicate unfavorable situations, as well as to facilitate proactive corrective actions for the best possible plant growth, is demonstrated by this visual representation, which is persuasive evidence of the system's capability quickly and precisely.

## 8.6 CONCLUSION AND FUTURE SCOPE

By utilizing the potential of Edge AI to maximize greenhouse plant development, this study represents a significant advance in the field of smart agriculture. The HTU21D humidity and temperature sensor, the OLED screen, and the STM32 Nucleo F401RE microcontroller have been successfully integrated with one another, demonstrating the potential of an intelligent system capable of analyzing and interpreting environmental parameters essential for plant growth.

The fact that this accomplishment was made possible despite scarce resources must be acknowledged. There is room for improvement and optimization with access to improved sensors, a more sophisticated display screen, and a quicker microcontroller. The potential that can be realized with more technological and material expenditure is demonstrated by this work.

In summary, this work serves as an example of how, even with limited resources, creative solutions may be created to tackle real-world problems. We provide the foundation for a more productive, environmentally friendly future by fusing technology and agricultural methods. Looking ahead, it is apparent that subsequent versions of this work will gain from additional cutting-edge components, ultimately resulting in improved greenhouse plant growth control.

## REFERENCES

[1] P. Kirci, E. Ozturk, Y. Celik, P. Kirci, E. Ozturk, and Y. Celik, "A Novel Approach for Monitoring of Smart Greenhouse and Flowerpot Parameters and Detection of Plant Growth with Sensors," *Agriculture 2022*, vol. 12, no. 10, p. 1705, Oct. 2022, doi: 10.3390/AGRICULTURE12101705

[2] C. J. Brien, B. Berger, H. Rabie, and M. Tester, "Accounting for variation in designing greenhouse experiments with special reference to greenhouses containing plants on conveyor systems," *Plant Methods*, vol. 9, no. 1, pp. 1–22, Feb. 2013, doi: 10.1186/1746-4811-9-5/FIGURES/9

[3] S. Vialet-Chabrand, J. S. A. Matthews, A. J. Simkin, C. A. Raines, and T. Lawson, "Importance of Fluctuations in Light on Plant Photosynthetic Acclimation," *Plant Physiol*, vol. 173, no. 4, p. 2163, Apr. 2017, doi: 10.1104/PP.16.01767

[4] N. Fatima, S. A. Siddiqui, and A. Ahmad, "IoT-based Smart Greenhouse with Disease Prediction using Deep Learning." *[Online]*. Available: www.ijacsa.thesai.org, doi: 10.14569/IJACSA.2021.0120713

[5] Christian von Zabeltitz, Integrated Greenhouse Systems for Mild Climates: Climate Conditions, Design, Construction, Maintenance, *Climate Control*. 2011, doi: https://doi.org/10.1007/978-3-642-14582-7

[6] M. R. A. Azhar, M. Hamid, M. H. Irfan, M. Awais, U. S. Khan and A. Zeb, "Automated greenhouse system," *2019 2nd International Conference on Communication, Computing and Digital systems (C-CODE)*, Islamabad, Pakistan, 2019, pp. 215–219, doi: 10.1109/C-CODE.2019.8681013

[7] T. Bhuvaneswari and J. T. H. Yao, "Automated greenhouse," *2014 IEEE International Symposium on Robotics and Manufacturing Automation, IEEE-ROMA2014*, pp. 194–199, Oct. 2015, doi: 10.1109/ROMA.2014.7295887

[8] H. Jaiswal, P. Karmali Radha, R. Singuluri, and S. A. Sampson, "IoT and Machine Learning based approach for Fully Automated Greenhouse," *2019 IEEE Bombay Section Signature Conference, IBSSC 2019*, vol. 2019 January, Jul. 2019, doi: 10.1109/IBSSC47189.2019.8973086

[9] "UK-RAS White papers The Future of Robotic Agriculture", *[Online]*. Available: www.ukras.org, (accessed Feb. 22, 2024).

[10] A. A. A. Halim, N. M. Hassan, A. Zakaria, L. M. Kamaruddin, and A. H. A. Bakar, "Automated scheduling based on plant growth for greenhouse management system," *2016 3rd International Conference on Electronic Design, ICED 2016*, pp. 230–235, Jan. 2017, doi: 10.1109/ICED.2016.7804643

[11] R. Zhao and L. Lu, "Automatic Temperature and Humidity Detection and Alarm System for Greenhouse," *IOP Conf Ser Earth Environ Sci*, vol. 512, no. 1, Jun. 2020, doi: 10.1088/1755-1315/512/1/012099

[12] K. Pallavi, J. D. Mallapur, and K. Y. Bendigeri, "Remote sensing and controlling of greenhouse agriculture parameters based on IoT," *2017 International Conference on Big Data, IoT and Data Science, BID 2017, vol. 2018-January*, pp. 44–48, Jul. 2017, doi: 10.1109/BID.2017.8336571

[13] P. Bhadani and V. Vashisht, "Soil moisture, temperature and humidity measurement using arduino," *Proceedings of the 9th International Conference On Cloud Computing, Data Science and Engineering, Confluence 2019*, pp. 567–571, Jan. 2019, doi: 10.1109/CONFLUENCE.2019.8776973

[14] F. A. Khan, A. A. Ibrahim, and A. M. Zeki, "Environmental monitoring and disease detection of plants in smart greenhouse using internet of things," *J Phys Commun*, vol. 4, no. 5, p. 055008, May 2020, doi: 10.1088/2399-6528/AB90C1

[15] L. Lipper *et al.*, "Climate-smart agriculture for food security," *Nature Climate Change 2014 4:12*, vol. 4, no. 12, pp. 1068–1072, Nov. 2014, doi: 10.1038/nclimate2437

[16] P. Tripicchio, M. Satler, G. Dabisias, E. Ruffaldi, and C. A. Avizzano, "Towards Smart Farming and Sustainable Agriculture with Drones," *Proceedings - 2015 International Conference on Intelligent Environments, IE 2015*, pp. 140–143, Aug. 2015, doi: 10.1109/IE.2015.29

[17] T. Ahonen, R. Virrankoski, and M. Elmusrati, "Greenhouse monitoring with wireless sensor network," *2008 IEEE/ASME International Conference on Mechatronics and Embedded Systems and Applications, MESA 2008*, pp. 403–408, 2008, doi: 10.1109/MESA.2008.4735744

[18] Z. Tafa, F. Ramadani, and B. Cakolli, "The design of a ZigBee-based greenhouse monitoring system," *2018 7th Mediterranean Conference on Embedded Computing, MECO 2018 - Including ECYPS 2018, Proceedings*, pp. 1–4, Jul. 2018, doi: 10.1109/MECO.2018.8405966

[19] H. Liu, Z. Meng, and S. Cui, "A wireless sensor network prototype for environmental monitoring in greenhouses," *2007 International Conference on Wireless Communications, Networking and Mobile Computing, WiCOM 2007*, pp. 2344–2347, 2007, doi: 10.1109/WICOM.2007.584

[20] M. Azaza, C. Tanougast, E. Fabrizio, and A. Mami, "Smart greenhouse fuzzy logic based control system enhanced with wireless data monitoring," *ISA Trans*, vol. 61, pp. 297–307, Mar. 2016, doi: 10.1016/J.ISATRA.2015.12.006

[21] T. Swathi, T. D. Rasiga, and S. Sudha, "IOT Based Agriculture Monitoring System," *Int J Res Appl Sci Eng Technol*, 2022, doi: 10.1109/ICICAT57735.2023.10263767

[22] "Preliminary infrastructure development for greenhouse accounting of Malaysian rainforest using wireless sensor network | Request PDF." Accessed: Feb. 15, 2024. [Online]. Available: https://www.researchgate.net/publication/279549048_Preliminary_infrastructure_development_for_greenhouse_accounting_of_Malaysian_rainforest_using_wireless_sensor_network

[23] A. Somov *et al.*, "Pervasive Agriculture: IoT-Enabled Greenhouse for Plant Growth Control," *IEEE Pervasive Comput*, vol. 17, no. 4, pp. 65–75, 2018, doi: 10.1109/MPRV.2018.2873849

[24] D. H. Park and J. W. Park, "Wireless Sensor Network-Based Greenhouse Environment Monitoring and Automatic Control System for Dew Condensation Prevention," *Sensors*, vol. 11, no. 4, pp. 3640–3651, Mar. 2011, doi: 10.3390/S110403640

[25] A. G. Q. Aquino, A. H. Ballado, and A. V. Bautista, "Implementing a Wireless Sensor Network with Multiple Arduino-Based Farming Multi-Sensor Tool to Monitor a Small Farm Area Using ESP32 Microcontroller Board," *2021 IEEE 13th International Conference on Humanoid, Nanotechnology, Information Technology, Communication and Control, Environment, and Management, HNICEM 2021*, 2021, doi: 10.1109/HNICEM54116.2021.9731989

[26] B. J. Kang, D. H. Park, K. R. Cho, C. S. Shin, S. E. Cho, and J. W. Park, "A study on the greenhouse auto control system based on wireless sensor network," *Proceedings - 2008 International Conference on Security Technology, SecTech 2008*, pp. 41–44, 2008, doi: 10.1109/SECTECH.2008.60

[27] E. Omrani, B. Khoshnevisan, S. Shamshirband, H. Saboohi, N. B. Anuar, and M. H. N. M. Nasir, "Potential of radial basis function-based support vector regression for apple disease detection," *Measurement*, vol. 55, pp. 512–519, Sep. 2014, doi: 10.1016/J.MEASUREMENT.2014.05.033

[28] H. Li, Y. Guo, H. Zhao, Y. Wang, and D. Chow, "Towards automated greenhouse: A state of the art review on greenhouse monitoring methods and technologies based on internet of things," *Comput Electron Agric*, vol. 191, p. 106558, Dec. 2021, doi: 10.1016/J.COMPAG.2021.106558

[29] "Climate change and food security: risks and responses.", https://www.fao.org/3/i5188e/I5188E.pdf (accessed Feb. 22, 2024).

[30] Y. Guo, H. Zhao, S. Zhang, Y. Wang, and D. Chow, "Modeling and optimization of environment in agricultural greenhouses for improving cleaner and sustainable crop production," *J Clean Prod*, vol. 285, p. 124843, Feb. 2021, doi: 10.1016/J.JCLEPRO.2020.124843

[31] E. Iddio, L. Wang, Y. Thomas, G. McMorrow, and A. Denzer, "Energy efficient operation and modeling for greenhouses: A literature review," *Renew Sustain Energy Rev*, vol. 117, p. 109480, Jan. 2020, doi: 10.1016/J.RSER.2019.109480
[32] L. K. S. Tolentino *et al.*, "Development of an IoT-based aquaponics monitoring and correction system with temperature-controlled greenhouse," *Proceedings - 2019 International SoC Design Conference, ISOCC 2019*, pp. 261–262, Oct. 2019, doi: 10.1109/ISOCC47750.2019.9027722
[33] V. P. Kumar, K. C. Ramya, J. S. Abishek, T. S. Arundhathy, B. Bhavvya, and V. Gayathri, "Smart garden monitoring and control system with sensor technology," *2021 3rd International Conference on Signal Processing and Communication, ICPSC 2021*, pp. 93–97, May 2021, doi: 10.1109/ICSPC51351.2021.9451788
[34] STMicroelectronics, 'STM32 Nucleo F401RE Datasheet.', https://www.st.com/en/evaluation-tools/nucleo-f401re.html (accessed Feb. 22, 2024).
[35] TEC, "HTU21D Datasheet", https://www.alldatasheet.com/datasheet-pdf/pdf/970004/HUMIREL/HTU21D.html (accessed Feb. 22, 2024).
[36] Waveshare, '1.3-inch OLED (B) Datasheet', https://www.waveshare.com/1.3inch-oled-b.htm (accessed Feb. 22, 2024).

Chapter 9

# Agricapital, a three-way agro-market aggregation using Cluster CRUD operations and role of ML in agriculture

*Shailendra S. Gaur, Sanjeev Kumar and Shweta Gupta*

## 9.1 INTRODUCTION

India's agrarian landscape is characterized by its immense contribution to the sustenance of over half of the nation's population. However, the agricultural sector encounters challenges, primarily marked by intermediaries that hinder seamless transactions between farmers and consumers. To rectify this, the agricapital platform emerges, serving as a transformative solution that seeks to establish a direct connection between farmers and consumers throughout India.

This document comprehensively addresses the technical requisites for the implementation of the designated problem statement, centered around a pioneering "Three Way Agro-Marketplace" christened agricapital. The operational framework of the project is elucidated through a blend of factual data, graphical representations, and diagrams. Complex technical jargon is meticulously deconstructed to offer a lucid depiction of this full-stack web application.

A. **Problem Statement:** Within India, where a substantial fraction of the economy hinges on agriculture, the imperative arises to cultivate a centralized marketplace catering to the entire spectrum of agricultural products and their derivatives. This endeavor holds the potential to be an indispensable asset for the agricultural ecosystem. Deconstructing the problem statement further, the application's ambit encompasses the establishment of a marketplace accommodating three distinct user categories:
   i. Farmer: Positioned as the optimal platform, this segment facilitates the procurement of farming-related requisites and facilitates the sale of agricultural yields.
   ii. Supplier: The platform offers a conduit to vend an extensive array of farming-related merchandise. Its expansive reach encompasses countless farmers spanning the entire nation.

DOI: 10.1201/9781003628392-9

iii. Consumer: A virtual realm wherein grains, vegetables, and an array of agro products are directly sourced from the diligent hands of farmers.

B. **Purpose:** At the heart of this full-stack application resides an unequivocal purpose: to orchestrate the creation of an all-encompassing marketplace, culminating in the eradication of intermediary entities bridging the gap between farmers and consumers. Simultaneously, it lays the groundwork for fostering direct commerce interactions between suppliers and farmers, a facet hitherto deficient. Through this innovative intervention, the application strives to bestow due advantages upon farmers who have remained underserved.

C. **Needs and Wants of Stakeholders:** The stakeholders in the Indian agriculture sector have distinct needs and wants that the proposed solution seeks to address:

i. Farmers:
- Need: Access to a direct and transparent marketplace where they can sell their products without intermediaries, ensuring fair prices and enhanced profits.
- Want: User-friendly tools and interfaces that enable easy listing of their products, real-time communication with buyers, and secure payment gateways.

ii. Consumers:
- Need: Access to a wide variety of fresh and high-quality agricultural products at reasonable prices.
- Want: A user-friendly platform with comprehensive product listings, transparent information about the origin of products, and a seamless purchasing experience.

iii. Suppliers:
- Need: A reliable channel to showcase and sell their agricultural products and services directly to farmers.
- Want: A platform that enables them to reach a vast network of farmers, manage their inventory efficiently, and receive prompt payments.

iv. Regulatory Authorities:
- Need: A system that ensures traceability, authenticity, and adherence to regulations in agricultural transactions.
- Want: Integration of standardized practices and compliance checks within the platform, aiding in oversight and regulatory enforcement.

v. Technology Developers:
- Need: A clear understanding of the technical requirements and user expectations to develop a robust and secure platform.
- Want: Access to necessary resources, collaboration with domain experts, and the opportunity to create innovative solutions that drive positive change.

vi. Investors:
- Need: Confidence in the viability and potential impact of the proposed solution in transforming the agricultural sector.
- Want: A clear roadmap of the project, transparent communication about progress, and a promising return on their investment.

In addressing these needs and wants, the proposed solution endeavors to revolutionize the Indian agriculture sector, fostering efficiency, transparency, and equitable growth opportunities.

## 9.2 PRODUCT DESCRIPTION

This document contains the problem statement that the current system is facing that is hampering the growth opportunities in Indian agriculture sector as shown in Figure 9.1. It further contains a list of the stakeholders and users of the proposed solution. It also illustrates the needs and wants of the stakeholders that were identified.

Machine learning (ML) models can analyze vast datasets related to crop health, climate conditions, and historical disease patterns. By identifying subtle correlations, these models can predict the onset of diseases, allowing farmers to take preventive measures and minimize crop losses. Real-time monitoring using sensors and satellite imagery, combined with ML algorithms, enables early detection of stress factors, facilitating timely interventions.

Automation has become a key focus in modern agriculture, and ML contributes significantly to automated harvesting and sorting processes.

*Figure 9.1* A section of home page.

Computer vision algorithms can be trained to recognize ripe fruits or vegetables, guiding robotic harvesters to pick crops with precision. This not only reduces the need for manual labor but also minimizes wastage by ensuring that only the ripest produce is harvested.

The impact of climate change on agriculture is a growing concern. ML models can analyze climate data to provide accurate predictions and assist farmers in adapting to changing conditions. By understanding climate patterns, farmers can adjust planting schedules, choose appropriate crops, and implement resilient farming practices. This proactive approach enhances the sector's resilience to the uncertainties posed by climate change.

ML contributes to optimizing the agricultural supply chain by enhancing logistics, storage, and distribution processes. Predictive analytics models can forecast demand based on various factors, helping farmers and distributors manage inventory efficiently. This results in reduced wastage, lower transportation costs, and improved overall supply chain efficiency.

As we stand on the cusp of a technological revolution, the integration of ML into agriculture holds immense potential. The benefits extend beyond mere efficiency gains; they include sustainable resource management, enhanced resilience to climate change, and the ability to feed a growing global population. By harnessing the power of algorithms and data analytics, farmers can usher in a new era of precision, sustainability, and productivity in agriculture. As we continue to explore the capabilities of ML, the future of farming looks promising, with technology playing a pivotal role in shaping a more sustainable and resilient agricultural landscape.

### 9.2.1 Software Requirements Specification (SRS)—Agricapital

A. **Restful Functionalities**: The foundational structure of agricapital, despite being a comprehensive full-stack application, is fundamentally rooted in its backend. This backend furnishes the essential CRUD (Create, Read, Update, Delete) functionalities, forming the bedrock of the system. The architecture adheres to the REST (Representational State Transfer) software style, defining the fundamental conventions for establishing endpoints within the application.
B. **Three-Way Agro-Market Aggregation**: Viewed from a holistic perspective, agricapital emerges as a tripartite market aggregation platform, catering prominently to three key user segments: farmers, consumers, and suppliers as shown in Figure 9.2. The platform is intricately designed to stimulate a hyperlocal and modular marketplace, fostering localized interactions and seamless transactions.

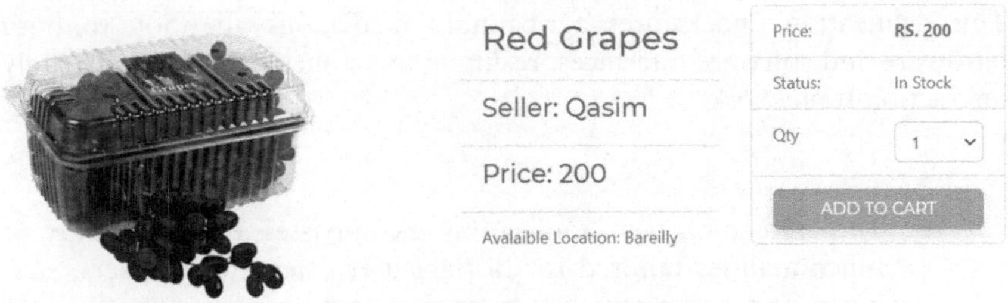


Figure 9.2 Snapshot of agro-product (grapes) sold by a farmer.

## 9.3 OVERALL DESCRIPTION

A. User Interfaces:
   i. Graphical User Interface (GUI): agricapital utilizes the Windows/Linux operating systems as its GUI, providing users with a familiar and intuitive environment.
   ii. Hardware Interface: The application is internet-dependent; thus, the primary hardware interface necessitates WLAN or Ethernet connectivity, facilitating internet access for the system.
   iii. Software Interface: agricapital interfaces with the following software components:
      a. Operating systems: Windows/Linux
      b. Programming languages: javascript
      c. Integrated development environment (IDE): Visual Studio Code (VS Code)
      d. Web browser: Google Chrome
   iv. Communication Interface: The interaction between clients and the server transpires through HTTPS (Hypertext Transfer Protocol Secure), ensuring secure and encrypted communication. Sign in page is clearly shown in Figure 9.3

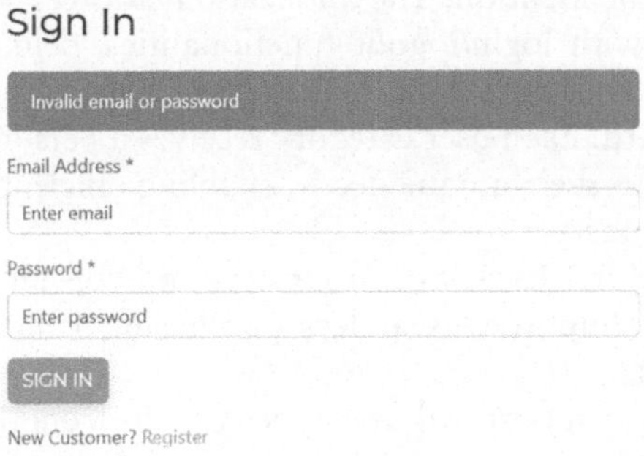


*Figure 9.3* Sign-in page.

This delineation underscores agricapital's meticulous attention to both hardware and software interfaces, resulting in a coherent and user-friendly interaction framework.

B. Product Features:
   - The agricapital web application encompasses a diverse array of functionalities tailored for distinct user categories: farmers, suppliers, and consumers. These features empower users to engage in specific actions aligned with their roles.

C. Farmer:
   - Purchase Essentials: Farmers can conveniently procure seeds, fertilizers, and machinery, streamlining their agricultural operations.
   - Sell Produce: Farmers possess the capability to market and vend their agricultural yields and materials directly on the platform.
   - Machinery Services: Farmers can offer heavy machinery and tractors for rent, fostering collaborative resource utilization.
   - Personalized Dashboard: A dedicated dashboard empowers farmers to efficiently manage and monitor their activities.

D. Supplier:
   - Machinery Sales: Suppliers are enabled to sell heavy machinery and equipment directly to farmers, enhancing accessibility to essential tools.
   - Supply Seeds and Fertilizers: Suppliers can provide farmers with seeds and fertilizers, facilitating the growth of agricultural endeavors.
   - Dashboard Management: Suppliers wield the authority to oversee and manage their distinctive dashboards.

E. Consumer:
   - Direct Produce Purchase: Consumers enjoy the privilege of procuring a diverse range of agricultural produce directly from farmers, eliminating intermediary involvement.
   - User Authentication: The application enforces secure user authentication with login/logout functionalities, delineating user access rights.
   - Dashboard: Each user category receives a personalized dashboard, tailored to their unique needs, enabling efficient oversight of their transactions.
   - Cart Feature: Exclusive to farmers and consumers, the cart functionality empowers seamless purchasing while not extended to suppliers.
   - Technical Support: All user types benefit from dedicated technical support to resolve any queries or confusion.
   - Payment Integration: Users are granted a convenient payment mechanism via the PayPal gateway.

F. Constraints:

- The application's mobile compatibility is effective, but it wasn't specifically optimized for smartphones.
- Establishing a successful marketplace mandates substantial groundwork, including the consolidation of suppliers and customers.
- The application's database is harnessed through no SQL databases, influencing its data management approach.
- Captcha functionality during the login process is not currently supported.
- The application's structure is rooted in user schema, product schema, and supplier schema for organized data management.

## 9.4 SPECIFIC REQUIREMENTS

### 9.4.1 Functional requirements

A. Frontend: the project extensively employs react.js, a javascript library renowned for building robust user interfaces. react.js structures user interfaces into reusable, independent components, promoting a separation of concerns. It leverages jsx, facilitating an enhanced approach to separation of UI elements.

B. Backend: node.js, a versatile javascript runtime, propels the backend of the application. This runtime enables javascript to function beyond the confines of the browser. Employed for its lightweight nature, node.js is complemented by express.js—a popular framework. express.js simplifies http request handling and the creation of tailored APIs, further fortifying the backend.

C. Database: the project relies on mongodb, a cross-platform document-oriented database program. Characterized as a no SQL database, mongodb employs json-like documents with the flexibility of optional schemas. mongodb synergizes seamlessly with node.js, delivering speed and robustness. The application further harnesses mongoose, an object modeling tool that streamlines scheme generation and query building.

D. Models: models wield a pivotal role in validating data integrity and ensuring the database's cleanliness. Through rigorous validation procedures, models assure that only accurate and consistent data populates the database, enhancing overall system reliability.

E. Controllers: controllers assume the responsibility of structuring api responses for frontend consumption. These entities encapsulate the core business logic pertinent to each endpoint. From data validation to processing intricate business rules, controllers culminate in the creation of a coherent api that either stores data or transmits it to the client.

F. Authentication: the implementation of robust authentication necessitates secure password storage, a feat accomplished through bcrypt—a widely adopted hashing function. bcrypt safeguards the application against potential breaches by fortifying password security.
G. Json web token (jwt): the application integrates json web tokens (jwt) to bolster security. jwt facilitates data creation, optionally with signatures and encryption. The payload, encompassing json data, asserts claims. This technology adds an extra layer of protection; tokens are signed using either private secrets or public/private key pairs.

### 9.4.2 Non-functional requirements

A. Efficiency requirement: the system is designed to ensure efficient data accessibility and modification for both farmers and consumers. Quick access to data is paramount, facilitating streamlined user interactions.
B. Reliability requirement: a fundamental aspect is the accurate execution of member registration and recognition within the system. The platform's reliability hinges on the seamless performance of these critical functions.
C. Usability requirement: the system embodies a user-centric ethos, prioritizing an environment characterized by user-friendliness. This design principle enables staff to effortlessly navigate and execute tasks. However, it's crucial to note that a constant internet connection is necessary for the proper functioning of the web application.
D. Performance requirement: optimized performance is observed when utilizing Google Chrome, leveraging its low storage and ram consumption. The system's performance is predominantly influenced by the speed of the internet connection. The application's efficiency is evident in the rapid display of currently listed camps on the homepage.
E. Reliability requirement (reiterated): the web application maintains its reliability, functioning robustly barring instances of unstable internet connectivity, which could potentially lead to temporary downtime.
F. Availability: the web application is readily accessible via standard web browsers, ensuring wide availability. Its user-friendly interface enhances user experience.
G. Security: the web application prioritizes stringent security parameters, safeguarding user data against potential threats. The development process meticulously accounts for security considerations.
H. Maintainability: effortless maintainability characterizes the web application, requiring minimal personnel for upkeep and continuous functioning.
I. Portability: the web application's inherent cross-browser compatibility ensures smooth portability. The ease with which it can be transitioned to alternative browsers underscores its versatility and accessibility.

### 9.4.3 Cluster CRUD Operations

1. Consumer-Product Controller:
   - Retrieve all products available for consumers.
   - Retrieve specific consumer products by their unique ID.
   - Delete a consumer product.
   - Facilitate the creation and updating of consumer profiles.
2. Order Controller:
   - Retrieve all orders associated with the logged-in user.
   - Initiate the creation of new orders.
   - Fetch an order using its unique ID.
   - Enable the update of order status to "paid" or "delivered."
3. Lending Controller:
   - Fetch all lending machines accessible.
   - Retrieve a specific lending machine based on its ID.
   - Facilitate the creation of a lending machine order.
   - Manage updates to an existing lending machine order.
4. Seed Controller:
   - Retrieve a comprehensive list of available seeds.
   - Delete a seed model.
   - Update seed model attributes.
   - Fetch specific seeds by their unique ID.
   - Establish the capability to create or update product reviews for seeds.
5. Supplier Controller:
   - Create a new supplier's product model.
   - Retrieve products associated with the logged-in supplier.
   - Fetch all products (admin functionality).
   - Update supplier profile information.
   - Enable updating of product reviews submitted by suppliers.
6. User Controller:
   - Implement authorization mechanisms using jwt.
   - Facilitate the registration of new users.
   - Retrieve and provide access to user profiles.
   - Update user profile information.
   - Allow for the deletion of user profiles.

## 9.5 SYSTEM DESIGN AND ANALYSIS

### 9.5.1 CRUD functionality in modern applications

In the contemporary landscape of software development, CRUD applications have emerged as a cornerstone. A CRUD application encapsulates four fundamental operations: "Create, Read, Update, Delete," forming the nucleus of its functionality. At its essence, a CRUD application comprises three integral components: the database, user interface, and apis. The orchestration of CRUD functionalities within this project is realized through the adoption of a restful architecture, culminating in the creation of REST apis.

### 9.5.2 Restful architecture and CRUD

The project seamlessly integrates CRUD functionalities through the prism of a restful architecture. As shown in the Figure 9.4 mentioning the Representational State Transfer (REST) APIs play a pivotal role in channeling these operations, facilitating an efficient communication between the user interface and the underlying database. REST APIs operate on the principles of simplicity, modularity, and ease of integration, enabling the project to adhere to the tenets of a modern and streamlined application design.

At the apex of the model, the user interface serves as the gateway through which users interact with the application's functionalities. The restful apis, situated in the middle tier, facilitate seamless communication between the user interface and the database. These apis expose endpoints corresponding to the CRUD operations, orchestrating the flow of data.

Incorporating the principles of REST, the project achieves a modular and well-structured design, enhancing maintainability and extensibility. This REST API model empowers the project to implement CRUD operations efficiently, providing a robust foundation for the application's core functionality as clearly mentioned in the Figure 9.5.

## 9.6 USE CASE DIAGRAM

The app serves three different kinds of users, and hence three use case diagrams are provided and shown in Figures 9.6 and 9.7.

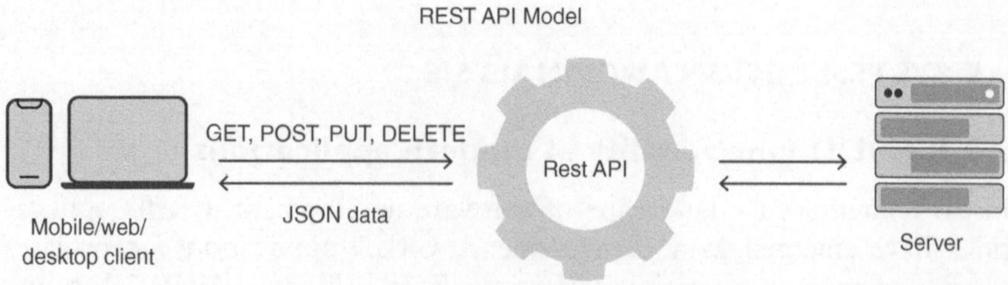


*Figure 9.4* Representational State Transfer (REST) API model.

```sql
+-----------------+
|  User Interface |
+-----------------+
        |
        v
+-----------------+
|  RESTful APIs   |
+-----------------+
        |
        v
+-----------------+
|    Database     |
+-----------------+
```

*Figure 9.5* RESTFUL API model.

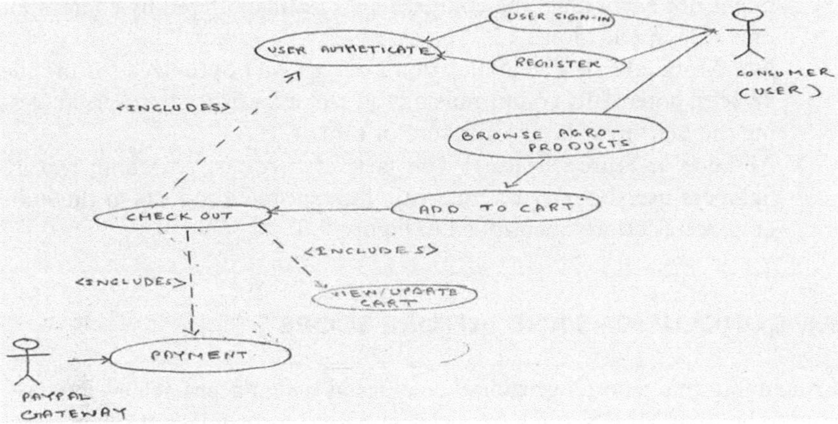


*Figure 9.6* Consumer use case diagram.

## 9.7 LIMITATIONS

1. Clunky User Experience: The user experience within the application can be improved to ensure smoother navigation and interactions, enhancing overall user satisfaction.
2. Limited Database Storage: The application's data storage capacity is constrained, potentially limiting its ability to accommodate extensive data and scalability.

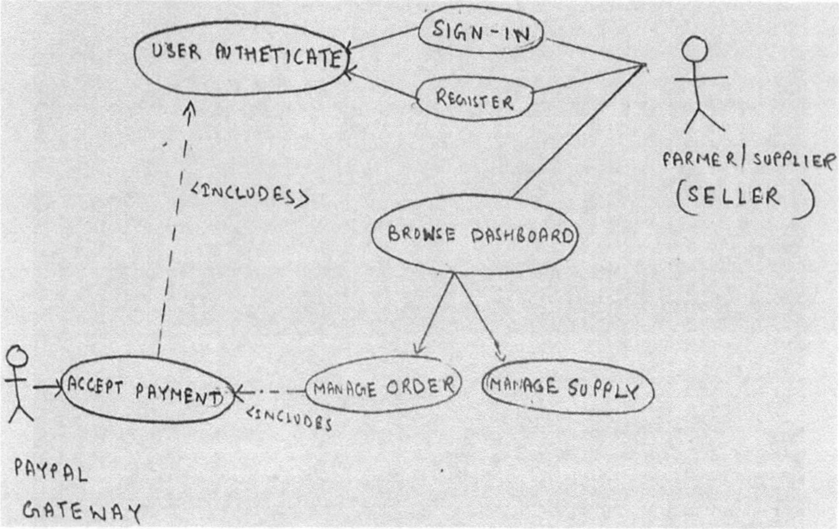


*Figure 9.7* Farmer/supplier use case diagram.

3. Feature Limitations: The application's feature set, while functional, might not encompass the comprehensive range offered by contemporary web applications.
4. Not Mobile-First: The application's design isn't optimized for mobile devices, potentially compromising the user experience for users accessing the platform via smartphones or tablets.
5. Absence of Search Feature: The lack of a robust searching feature hampers users' ability to efficiently find specific products in the marketplace as clearly mentioned in Figure 9.8.

## 9.8 CONCLUSIONS AND FUTURE SCOPE

Concluding this report, agricapital emerges as a significant achievement—a comprehensive MERN stack application catering to India's thriving agro-market. A MERN stack is a java script stack that is used for deployment of full stack web applications. The project has not only enriched the knowledge of full-stack web development but has also provided valuable insights into India's agri-tech industry.

**Future Scope:** As with any evolving application, agricapital has potential avenues for enhancement and refinement. Some key areas for improvement include:

- Two-Factor Authentication: Bolstering security through two-factor authentication would be a valuable addition, safeguarding user accounts.

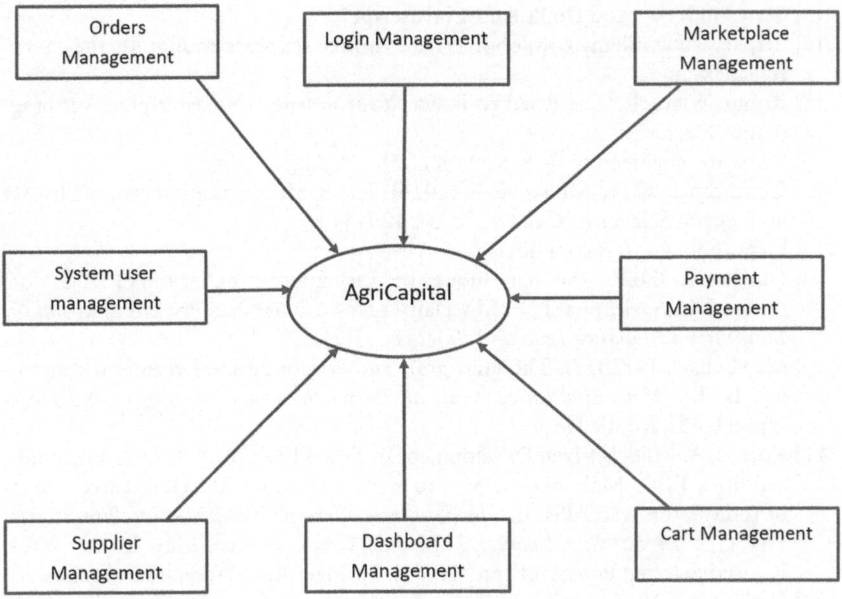


Figure 9.8 DFD Level 0.

- Enhanced User Experience/Interface: Continual refinement of the user experience and interface holds promise for optimizing user engagement.
- Integration of Captcha: Integrating captchas into login and registration forms can enhance security against automated attacks.
- Search and Sorting Feature: Introducing a robust search and sorting feature for agricultural products would significantly enhance user navigation and product discovery.
- Wishlist Functionality: Integrating a "wishlist" feature aligns with marketplace norms, enabling users to save and track products of interest.
- Recommendation Logic: Implementing recommendation logic can elevate user experience by providing personalized product suggestions.

The journey of agricapital doesn't conclude here—it sets the stage for future enhancements and innovations, reaffirming its potential impact on India's agricultural sector.

## BIBLIOGRAPHY

[1] IEEE Recommended Practice for Software Requirements Specifications-IEEE Std. 830–1993.
[2] https://reactjs.org/docs/state-and-lifecycle.html

[3] Simpson Kyle "You Don't Know javascript"
[4] https://www.udemy.com/course/react-nodejs-express-mongodb-the-mern-fullstack-guide
[5] Robin, Wieruch "The Road to React: Your journey to master plain yet pragmatic React.js".
[6] Software Engineering by K.K. Aggarwal and Yogesh Singh.
[7] Clutterbuck, C., & Millstone, E. (2016). The BSSRS agricapital group: History and legacy. *Science as Culture*, 25(3), 400–414.
[8] https://mongoosejs.com/docs/
[9] Duggan, A. (2020). Purchase money security interests under the PPSA: A note on'StockCoAgricapital Pty Ltd v Dairy Livestock Services Pty Ltd'. *Journal Of Banking and Finance Law and Practice*, 31(2), 170–175.
[10] McMichael, P. (2013). The land grab and corporate food regime restructuring. In *The New Enclosures: Critical Perspectives on Corporate Land Deals* (pp. 63–83). Routledge.
[11] Korzec, A. Model-driven Development of Front-Ends for CRUD Applications.
[12] Bülthoff, F., & Maleshkova, M. (2014). RESTful or RESTless–current state of today's top web APIs. In *The Semantic Web: ESWC 2014 Satellite Events: ESWC 2014 Satellite Events*, Anissaras, Crete, Greece, May 25–29, 2014, Revised Selected Papers 11 (pp. 64–74). Springer International Publishing.
[13] Shylesh, S. (2017, April). A study of software development life cycle process models. In *National Conference on Reinventing Opportunities in Management, IT, and Social Sciences* (pp. 534–541).

Chapter 10

# Optimizing smart and sustainable agriculture for sustainability

*S. Prabhakara*

## 10.1 INTRODUCTION

One of the important sectors that are essential to a nation's economic development is agriculture. It satisfies everyone's demand for food, around the world. The increase in various agricultural operations' capacity and efficiency has surely been aided by innovative agricultural improvements. There is ongoing pressure on the current global agriculture system to produce more food. The demand for this is primarily due to population expansion. The global population is projected to touch 8.5 billion people by 2030 and 9.6 billion by 2050, as per the estimates from the Food and Agriculture Organisation (FAO) of the United Nations (UN). This would lead to an unparalleled demand for food and agricultural products. To satisfy such a demand, food production is estimated to need to rise by 70 per cent [1–4]. To address the scarcity of water, the changing climate, the limited amount of land available for farming, and the frequently fluctuating environmental conditions, smart and sustainable agriculture measures must be created. To maintain food security, there has generally been an increase in the quantity of agricultural production over the years. Various initiatives are being initiated to enhance the quantity and quality of agricultural products by converting them into smart and associated products through a smart and sustainable way.

## 10.2 CONCEPT OF SUSTAINABILITY AND SUSTAINABILITY IN AGRICULTURE

Sustainability is a systems issue and refers to the eternal ability. As a goal, sustainability aims to make an activity or system permanent. A system is made up of various parts, pieces, subsystems, or constituents that work together to form an integrated whole that supports the system's structure and integrity maintenance [5]. According to Constanza and Palten (1995), a sustainable system is one that endures or perseveres despite new stressors and shocks because of the interdependent nature of its parts. Sustainability

DOI: 10.1201/9781003628392-10

does not imply that a system has an endless lifespan. A system is considered sustainable when it achieves its intended life period in a way that is consistent with its temporal and spatial scale [6].

Marcus Terentius Varro, a Roman landowner from the first century BC, deserves recognition for being the one who initially recognised and articulated the significance of sustainability in agriculture. Varro states that agriculture is a science that tells us what crops to grow on different types of soils and what practices to follow so that the land can always yield the most [7]. This is arguably the best and most concise explanation of the idea of sustainable agriculture that has been provided to date, and it serves as the foundation for most further expositions on the subject.

Redesigning the current agricultural development model is necessary to achieve the diametrically opposed goals of sustainability (environmental) and development (economic) at the same time [8]. The best way to comprehend the current state of agricultural sustainability is to trace the coevolution of production systems and agricultural practices with human needs and inventiveness. During the pre-industrial era for around 6000 years, from 8000 to 2000 BC, subsistence agriculture was practiced with little to no adverse impact on the environment. Early shifting cultivation methods and the conventional agricultural systems of ancient India, China, the Middle East, and other civilisations endured for thousands of years and produced enough food to sustain the low population at the time [8].

## 10.3 SUSTAINABLE DEVELOPMENT

Our Common Future (also known as the Brundtland Commission Report) defined sustainable development as "development that meets the needs of the present without compromising the ability of future generations to meet their own needs." It is a process in which the use of resources, the direction of investments, the alignment of technological development, and institutional change are all in harmony and endorse both the current and future potential to meet human needs and operations [9]. The world community was made aware of the serious environmental degradation, particularly in industrialised nations, by the report. Because the global environment has a limited ability to produce resources and absorb waste, unchecked growth could compromise the likelihood that coming generations will be able to meet their demands, making this degradation unsustainable.

The Brundtland Commission promoted an alternative paradigm of all-around, sustainable development. This is to ensure economic progress in addition to the safety of the biophysical environment, social equality, and harmony, so that a minimum standard of living became the right of all people [8]. Therefore, sustainable development can be defined as the peaceful rebalancing, within the environmental bounds of our planet, of economic growth, environmental safety, and the social well-being of all communities worldwide. It is commonly acknowledged that the concept of sustainable development needs to be at the heart of all upcoming human undertakings. The idea, to put it simply, combines two fundamental ideas: ecological sustainability and economic development [10]. A system or activity is not considered sustainable development if it does not incorporate these two concepts. Sustainable agriculture has gained prominence ever since the World Conference on Environment and Development (WCED) emphasised sustainable development as a paradigm for intergenerational equity [9].

## 10.4 SUSTAINABLE AGRICULTURE

The Technical Advisory Committee (TAC) of the Consultative Group on International Agricultural Research (CGIAR), FAO, UN defined sustainable agriculture as "the successful management of resources for agriculture to satisfy changing human needs while maintaining or enhancing the quality of the environment and conserving natural resources" [11]. Since increasing production and productivity was the primary goal of the CGIAR, the economic and social dimensions of sustainable agriculture are not explicitly included in the definition. "A sustainable agriculture is one that, over the long term, enhances environmental quality and the resource base on which agriculture depends; provides for basic human food and fibre needs, is economically viable; and enhances the quality of life for farmers and society as a whole," according to the American Society of Agronomy, which adopted a more holistic definition of the term in 1989 [8]. Thus, sustainable agriculture functions within the constraints of both socioeconomic viability and quality and physical and biological resources. It will need first and foremost to consider two inseparables, intertwined societal priorities-preserving the environment and providing safe and healthy food for all [12]. Gupta, Niti et al. (2021) identified 30 sustainable agriculture practices widespread in India [13]. "Sustainable agriculture includes adaptive measures taken to address the adverse effect of climate and enhance the resilience of the agricultural system" [2].

The notion of sustainable agriculture is best summed up by G.K. Douglas (1984) as "long-term food sufficiency which requires that agricultural systems be more ecologically based and do not destroy their natural base. Sustainability as stewardship means that agricultural systems are based on a conscious ethics regarding humankind's relationship to future generations and to other species in nature. Sustainability as a community means that agricultural systems are equitable. Agricultural systems cannot be sustained if there is gross misdistribution of land, wealth and power. Social tensions and upheavals in many parts of the world are often manipulations of such inequalities" [8]. According to a different definition provided by Mac Rae et al. (1990), sustainable agriculture is both an agricultural method and a philosophy. Its foundations are a collection of principles that demonstrate a commitment to correctly responding to the awareness of ecological and social realities. It places a strong emphasis on management practices and design that complement natural processes to preserve resources, reduce waste, and protect the environment all while preserving and enhancing agricultural output. The various facets of sustainability – physical, economic, ecological, social, cultural, and ethical – must be acknowledged in any definition. Only after defining what is to be sustained – that is, within the confines of a system's framework – can sustainability be defined [14]. Agricultural systems function at various levels, including soil-plant systems, cropping systems, farming systems, agro-ecosystems, and higher regional, national, and global levels, making boundary selection challenging [8]. Thus, the level selected also establishes the definition's spatial scale of action. Agro-ecosystem and higher levels are impacted by decisions made at the farm level, and vice versa. Important connections exist across agricultural systems at various spatial dimensions, or hierarchical levels [15]. Although the emphasis in each definition varies, they all emphasise the necessity for agricultural practices to be ecologically friendly, economically feasible, and considerate of human well-being to meet the world's food demands. Since there are several approaches to attain sustainability goals instead of just organic farming, agriculture is not associated with any specific technological technique. Agriculture is considered sustainable when it can buffer changes in the resource base and demand over time. Technology plays a crucial role in facilitating this process by providing time that is conducive to sustainability. The development of sustainable agricultural technology in response to the changing demands of society and the environment is guided by the concepts of conservation, complementarities, and resource rejuvenation. According to Rao (2002), the food demand of the expanding population and economy (sustainability goals) and the supply constraints imposed by the agro-ecosystem's carrying capabilities (system capacities) are the main factors that determine the framework of sustainable agriculture. 2. The trade-offs, as determined by trends in appropriate sustainability indicators, between the quality of the natural resource base and agricultural productivity in various regions/agro-ecosystems (Are the levels of and increase in production sustainable?), and 3. Developing technologies and better management techniques that can change the trade-offs to increase productivity and sustainability. (Can new technologies and management techniques improve the outlook for long-term sustainability?) [15]. The crop production system and the agroecosystem levels are the two levels at which the framework is to be implemented. Agriculture needs to be lucrative in the near term to be sustainable. Thus, it is also necessary to have two level perspectives: the short term and the long term. In contrast to the indicators of ecosystem health, which take longer to become effective, the sustainability indicators of profitability and social concerns based on income production and distribution work quickly.

## 10.5 SMART AGRICULTURE

Bach and Mauser (2018) concluded that smart farming, which uses earth observation data in farming, helps environmentally and financially sound agricultural management through site specific application [16]. This is a significant step towards sustainable agriculture. Smart and precision agriculture also known as smart farming is nothing but the combination of technology, such as the Internet of Things (IoT), sensors, robotics, artificial intelligence (AI), smart supply chains, big data analytics, and block chain, into the agriculture [17]. The phrases "smart farming," "agriculture 4.0," and "digital agriculture" are used to characterise the trend of more data-related technology being employed in farming and food production. These words are generally interchangeable [18].

By collecting real-time data on factors like soil moisture, temperature, crop growth, and livestock health, smart farming enables farmers to make informed decisions and take precise actions, leading to enhanced efficiency, reduced costs, and improved yields. With the aid of sensors and machine-based irrigation techniques, farmers can use digital applications to monitor the temperature, humidity, and moisture content of their agricultural area. If there is an adequate or inadequate moisture deficit on his/her farm, the farmer will receive an alarm on his/her cell phone. With sensors, he/she can keep an eye on the amount of water, fertilisers and pesticides needed. Bhuvan, an ISRO (Indian Space Research Organisation) geo platform,[1] offers useful information on weather, pest monitoring, and plantations, which helps reduce labour costs [19].

It is a solution-centred concept that gives the agriculture business with a progressive technology set up that uses big data, cloud computing, IoT for recognisability, robotising, and exploring various tasks. Further, it could be considered as a method to understand essential requirements as well as the adjustments in current climate settings that change sensor activities and the tasks of actuators. It is necessary to use equipment and programming innovations such as conveying sensor hubs, control frameworks, mechanical technologies, satellites for taking pictures and situating, information storage and examination, warning frameworks, drones, and robots [17].

The important components of smart farming include (1) Global Positioning System (GPS), (2) sensing and monitoring systems, (3) data analytics and predictive models, (4) automated systems and robotics, and (5) precision application techniques [17, 20].

*GPS*: Satellites of the GPS send out signals that enable GPS receivers to calculate their current location in real time and offer continuous positions even when they are moving. Farmers may find out the location of field data, including pest occurrence, soil type, weeds, and other obstacles, thanks to the accurate location information.

*Sensing and monitoring systems*: These employ a network of sensors and IoT devices to gather data on crucial agricultural parameters. Soil moisture sensors, weather stations, and aerial drones equipped with high-resolution cameras are some examples of these devices. They provide valuable insights into soil conditions, weather patterns, pest infestations, and other vital information necessary for effective farm management.

*Data analytics and predictive models*: The vast amount of data collected from sensing and monitoring systems would be of little use without powerful data analytics and predictive models. With the help of AI algorithms and machine learning techniques, farmers can analyse complex datasets and derive meaningful patterns and predictions. These models can provide recommendations on optimal irrigation schedules, disease prevention strategies, and even predict market demand, enabling farmers to make informed decisions and mitigate risks.

*Automated systems and robotics*: Automation plays a pivotal role in smart farming, streamlining labour-intensive tasks and reducing human error. Automated systems and robotics are utilised for activities such as planting, harvesting, and livestock management. For instance, self-driving tractors equipped with GPS technology can precisely sow seeds and apply fertilisers, while robotic milkers ensure efficient and stress-free milking processes. By automating repetitive tasks, farmers can save time, reduce costs, and allocate their resources more effectively.

*Precision application techniques*: These enhance resource efficiency and minimise environmental impact. Through targeted delivery of fertilisers, pesticides, and water, farmers can optimise input usage and reduce chemical runoff. Variable-rate technology enables the application of inputs based on specific crop needs, determined by analysing data collected from sensors and predictive models. By tailoring the application of resources, smart farming promotes sustainable practices and minimises waste.

## 10.6 OPTIMISING SMART AND SUSTAINABLE AGRICULTURE FOR ACHIEVING SUSTAINABILITY

It is believed that smart farming, both in crop production and in livestock farming, can lessen the ecological footprints of agriculture and ameliorate greenhouse gas emissions with relation to climate change and sustainability [21]. Furthermore, it is believed that smart farming might increase farmers' profits by avoiding risks and limiting resource usage using high-precision geographic data [22]. Farmers may be able to produce in a more profitable, resource-efficient, and sustainable manner that also improves the environment and climate with the help of smart farming and smart products [23]. By utilising a variety of technologies, crop and livestock production techniques, and networks including all stakeholders in the agri-food industry, smart farming provides a route toward sustainable agriculture [24]. Balducci F. et al. (2018) discussed smart farms models for a sustainable and optimised agriculture using different datasets [25].

Bineeta Satpathy (2022) discussed the examples such as the Association for People of Haryana (AFPOH), Kenya Agricultural and Livestock Research Organization (KALRO) (https://www.kalro.org), E-Krishok (http://www.krishok.com) and Zero Hunger in Bangladesh, and "Farmers" Advisory Information System in Tanzania for achieving sustainability through digital farming across the globe [19]. It is anticipated that smart farming will also increase consumer acceptance of agricultural products. A better-quality product may result from using fewer pesticides and fertilisers, which could raise higher price for produce. At the same time, digital tracking of agricultural products would allow for production transparency [26].

The IoT technology can be utilised in a variety of ways by agricultural organisations that specialise in outdoor farming to automate processes and maximise output. The need of various farming technologies, especially the IoT, was highlighted by Dhanaraju, M. et al. (2022) to make agriculture more intelligent and efficient to fulfil future demands. Variable-rate seeding is the process of precisely and correctly controlling where seeds are placed using modern farm equipment networked with wireless communications and GPS capabilities. IoT can combine technological developments in sensing and connectivity to transform agriculture into precision agriculture that is favourable to the environment [20].

Smart agriculture may incorporate overseeing effective resource management computerisation of repetitive tasks, and the use of big data, satellite navigation, sensor networks, and other technologies to find useful answers through data analytics. Farmers can now carry out tasks including monitoring, water use, satellite imagery, harvesting of crops, management of crops, optimisation, food security, control of temperature, and observation of soil quality thanks to smart agriculture and sensors. Further, sensors are applied to measure air and soil temperature, ambient light, moisture content and pH value of soil, velocity, and other parameters. The proper amount of water should be supplied by an effective irrigation system without wasting any. Therefore, it is important to accurately estimate how much water each location will require. According to Canavari et al. (2021), precision agriculture technologies possess the capability to alleviate the depletion of water resources that is frequently associated with traditional agricultural methods [27].

The excessive reliance on agrochemicals in modern agriculture has resulted in serious disruptions to agricultural production, variations in the amount and quality of output, harm to the environment, and serious financial and health hazards for farmers, and they need systems to diagnose and treat crop diseases [28]. Cutting-edge computer and network technology offer options for making wise judgements in applying fertiliser and moving animals to more practical and appropriate locations.

While utilising adaptable gadgets and other devices such as mobiles, smart agriculture includes fixing a farm or field by creating limits that consider the soil's moisture content, compost or other materials, environment, climate, and the creation of the required yields. As a result, data can be obtained from small holder farmers for digitisation. The data can be utilised to build profiles of farmers, which can then be improved over time by a variety of specialised organisations, including input suppliers, banks, and processors. Farmers can make effective use of the data and service providers, which will save expenses, improve returns, and create healthier farms. Sensors, cameras, and other devices are used by IoT in smart agriculture to gather data and make inferences. All devices are essentially connected to the Internet, creating a seamless ecosystem. In smart agriculture, where sensors offer data about fields and crops, IoT devices are essential.

To improve crop observation, cultivation, water systems, and herbicide and pesticide action, agriculture uses AI-based automated robotic systems. Smart machines have been equipped with AI to carry out daily chores. Advances in computer-based intelligence are being created to increase efficiency and assist in resolving several issues in the farming sector, such as crop yields and soil health. A system for managing the agriculture supply chain includes all the steps from the field or farm to the end user. It is a network of interconnected structures that relies on the greatest possible product generation and use. Great technologies like blockchain, AI, and the IoT are important for preserving an efficient food and agricultural supply chain system that ensures freshness of products, transparency in data, and increased consumer confidence [3].

Many of the problems facing conventional agriculture are resolved by the integration of wireless sensors and the IoT in smart farming. These problems include pest and disease control, irrigation, drought monitoring, land suitability through soil mapping and plant monitoring, yield maximisation through site-specific nutrient management, and yield monitoring and forecasting [20]. Sustainable precision agriculture and environment will contribute to enhanced yields and profits, improved adaptation to a fluctuating climate, enhanced sustainability of agricultural systems, and enhanced sustainability outside of the field and across watersheds, reducing nutrient transport across watersheds and adding to global sustainability [26], Furthermore, the benefits of smart farming for the environment also benefit society as a whole [29].

## 10.7 CHALLENGES OF SMART AGRICULTURE

The smart agriculture has the prospective to fully transform the agricultural sector, but to reap all its benefits, several issues need to be resolved. The following are a few of the main obstacles [20, 30, 31]:

1. *Cost*: The expense of integrating smart agriculture technologies is one of the main obstacles. A lot of technologies can be costly to buy and operate, like IoT, sensors, and precision farming tools. For many farmers, particularly small-scale growers in developing nations, this can be a hurdle to adoption.
2. *Management of data*: Large volume of data is generated by the newest agri-tech technologies which can make them challenging to handle and evaluate. The infrastructure and knowledge needed by farmers and agricultural organisations to manage the data and transform it into insights can be put into practice.
3. *Connectivity*: To transfer data, several technologies, like Wi-Fi and cellular networks, depend on connectivity. The efficacy of these technologies may be constrained, though, as many rural locations require more dependable connectivity.
4. *Lack of integration between the systems*: One area where smart farming technologies need to be further developed is integration across systems, which includes production, property management, and decision-making tools. The divide between information science and agriculture must be bridged via multidisciplinary group communication and academic exchange.
5. *Privacy and security*: Technologies create privacy and security issues, just like any other technology that produces data. Data security and individual privacy must be prioritised by farmers and agricultural organisations.
6. *Education and training*: To use the newest technology in the agri-tech sector efficiently, one needs to possess certain knowledge and training. To use the technologies and understand the data they provide, farmers and agricultural organisations need to be trained.

## 10.8 CONCLUSION

By integrating the power of technology, data analytics, and automation to create a sustainable and productive farming ecosystem, smart farming offers a major advancement in agricultural methods. Farmers may optimise productivity, reduce expenses, and minimise environmental effect by utilising automated systems, data analytics, sensing and monitoring systems, and precision application techniques. Smart farming is more than just an idea; it is a force for change that enables farmers to welcome innovation and ensure that agriculture has a bright future. One needs to take advantage of smart farming's opportunities and join the drive for a more productive and sustainable farming sector. To put it briefly, smart farming has changed the game in the agricultural sector by providing a host of benefits like higher yield, resource efficiency, environmental sustainability, and better decision making. Farmers may reduce their environmental impact and fully utilise their land by implementing smart agricultural practices.

## NOTE

1 Refer the link: https://bhuvan.nrsc.gov.in//home/index.php for further details.

## REFERENCES

1. Chehria, Abdellah, Chaibib, Hasna, Saadaneb, Rachid, Hakemc, Nadir, Wahbib, Mohamed, 2020, "A Framework of Optimizing the Deployment of IoT for Precision Agriculture Industry" in *24th International Conference on Knowledge-Based and Intelligent Information & Engineering Systems, Procedia Computer Science*, vol 176,2020, pp.2414–2422, https://doi.org/10.1016/j.procs.2020.09.312
2. Abdulla, N., Vasylieva, N., Volovyk, I. "Production Optimization for Sustainable Agriculture and Efficient Contract Farming in the Republic of Maldives", *Bulgarian Journal of Agricultural Science*, vol. 28, no.4, pp. 579–590, 2022
3. Jararweh, Yaser, Fatima, Sana, Jarrah, Moath, AlZu'bi, Shadi. "Smart and sustainable agriculture: Fundamentals, enabling technologies, and future directions, *Computers and Electrical Engineering*", vol. 110, 2023 available: https://doi.org/10.1016/j.compeleceng.2023.108799. (https://www.sciencedirect.com/science/article/pii/S0045790623002239)
4. Chaudhary, B., Kumar, V. "Emerging Technological Frameworks for the Sustainable Agriculture and Environmental Management", *Sustainable Horizons*, vol. 3, pp. 1–14.1995. https://doi.org/10.1016/j.horiz.2022.100026
5. Von Bertalanffy, L. *General System Theory: Foundations, Development, Applications*. New York. George Braziller, 1968.
6. Costanza, R. and Palten, B.C. "Defining and predicting sustainability." *Ecological Economics*, vol. 15, pp.193–196, 1995. https://doi.org/10.1016/0921-8009(95)00048-8
7. Conway, G.R. *The Doubly Green Revolution: Food for All in the 21st Century*. London: Penguin Books. 1997.
8. Hegde, D.M., Sudhakara Babu, S.N. (2016, July) "Sustainable Agriculture", *ISA article*, pp.1–41, https://www.scribd.com/document/690976524/Sustainable-Agriculture-ISA-article-DM-Hegde-SNS-Babu-July-30-2016
9. WCED. *Our Common Future*. Oxford University Press, Oxford, 1987
10. Braat, L. "The predictive meaning of sustainability indicators." In: *Search of Indicators of Sustainable Development* (O. Kurk and H. Vergruggen, Eds.). Dordrecht Kluwer Academic Publishers, The Netherlands, 1991. Environment & Management, vol 1. Springer, Dordrecht. https://doi.org/10.1007/978-94-011-3246-6_6
11. TAC. "*Sustainable Agricultural Production: Implications for International Agricultural Research*." Consultative Group for International Agricultural Research, Washington DC, March, 1988. https://hdl.handle.net/10947/1347

12. Oberč, B.P., Arroyo Schnell, A. "*Approaches to sustainable agriculture. Exploring the pathways towards the future of farming*". Brussels, Belgium: IUCN EURO, 2020. https://doi.org/10.2305/IUCN.CH.2020.07.en
13. Gupta, Niti, Shanal Pradhan, Abhishek Jain, Nahya Patel. *"Sustainable Agriculture in India" in What We Know and How to Scale Up*. New Delhi: Council on Energy, Environment and Water, 2021.
14. Mac Rae, R.J., Hill, S.B., Hennings, J. and Bentley, A.J. "Policies, programs and regulations to support the transition to sustainable agriculture in Canada" *American Journal of Alternative Agriculture*, vol.5, pp.76–92, 1990, https://doi.org/10.1017/S0889189300003325
15. Rao, N.H. "Sustainable agriculture: critical challenges facing the structure and function of agricultural research and education in India." In *National Workshop on Agricultural Policy: Redesigning R&D to Achieve the Objectives*, April, New Delhi, 2002.
16. Bach, H., Mauser, W. "Sustainable Agriculture and Smart Farming." In: Mathieu, P.P., Aubrecht, C. (eds) *Earth Observation Open Science and Innovation. ISSI Scientific Report Series*, vol. 15. 2018, Springer, Cham. https://doi.org/10.1007/978-3-319-65633-5_12, (http://creativecommons.org/licenses/by/4.0/)
17. Shet, Jayprakash. "Smart Farming: Revolutionizing Agriculture for a Sustainable Future," 2023, available: https://www.linkedin.com/pulse/smart-farming-revolutionizing-agriculture-sustainable-jayprakash-shet/ accessed on 15/11/2023
18. Hackfort, S. "Patterns of Inequalities in Digital Agriculture: A Systematic Literature Review." *Sustainability*, vol. 13, 12345. pp.1–18,2021. https://doi.org/10.3390/su132212345
19. Satpathy, Bineeta. "Digital transformation for sustainable agriculture: a progressive method for smallholder farmers," *Current Science*, vol.123 no.12, Dec. 2022, pp.1436–1440, https://www.currentscience.ac.in/Volumes/123/12/1436.pdf
20. Dhanaraju, M., Chenniappan, P., Ramalingam, K., Pazhanivelan, S., Kaliaperumal, R. "Smart Farming: Internet of Things (IoT)-Based Sustainable Agriculture." *Agriculture*, vol. 12, 1745, pp.1–126, 2022. https://doi.org/10.3390/agriculture12101745
21. Schulze, E., Luyssaert, S., Ciais, P. et al. "Importance of methane and nitrous oxide for Europe's terrestrial greenhouse-gas balance." *Nature Geo. Sci.*, vol. 2, pp. 842–850, 2009. Available: https://doi.org/10.1038/ngeo686
22. Power Alison, G. "Ecosystem services and agriculture: trade-offs and synergies," *Phil. Trans. R. Soc.*, 365, pp. 2959–2971, 2010 available: http://doi.org/10.1098/rstb.2010.0143
23. Schukat, S., Heise, H. "Towards an Understanding of the Behavioral Intentions and Actual Use of Smart Products among German Farmers.' *Sustainability*, 13, 6666. 2021, https://doi.org/10.3390/su131266 https://doi.org/10.3390/su13126666
24. Walter, Achim, Finger, Robert, Huber, Robert and Buchmann, Nina. "Smart farming is key to developing sustainable agriculture", *PNAS*, vol.114, no. 24, pp.6148–6150, 2017, https://doi.org/10.1073/pnas.1707462114
25. Balducci, F., Fomarelli, D., Impedovo, D., Longo, A., Pirlo, G. "Smart Farms for a Sustainable and Optimized Model of Agriculture" in 2018 *AEIT International Annual Conference*, Bari, 2018, pp.1–6. DOI: 10.23919/AEIT.2018.8577226

26. DeFries, R., Fanzo, J., Remans, R., Palm, C., Wood, S., Anderman, T.L. "Global nutrition. Metrics for land-scarce agriculture," *Science*, vol. 349, pp. 238–240, 2015.
27. Canavari, M., Medici, M., Wongprawmas, R., Xhakollari, V., Russo, S. A. Path Model of the Intention to Adopt Variable Rate Irrigation in Northeast Italy. *Sustainability*, vol. 13, 1879, 2021. https://doi.org/10.3390/su13041879
28. Sivagnanasundaram, J., Goonetillake, J., Buhary, R., Dharmawardhana, T., Weerakkody, R., Gunapala, D., Ginige, A. "Digitally-Enabled Crop Disorder Management Process Based on Farmer Empowerment for Improved Outcomes: A Case Study from Sri Lanka. *Sustainability*," 13, 7823, 2021, available: https://doi.org/10.3390/su13147823
29. Lieder, S. and Schröter-Schlaack, C. "Smart Farming Technologies in Arable Farming: Towards a Holistic Assessment of Opportunities and Risks." *Sustainability*, vol. 13, 6783.pp. 1–20, 2021. https://doi.org/10.3390/su13 126783
30. Gamage, T.A., Ranasinghe, D.D.M. "Optimizing Smart Agriculture by Integrating Artificial Cognition: A Review, Current Challenges and Future Trends," *International Journal of Research in Computing, (2820-2139)*, vol. 02, no. 1, pp. 11–20, July 2023.
31. Rahul, A. R. (2023, March). "Smart Agriculture: The Next Frontier in Sustainable Farming" https://blog.accubits.com/smart-agriculture-the-next-frontier-in-sustainable-farming/ accessed 15/11/2023.
32. Delgado, J.A., Short, N.M. Jr, Roberts, D.P., and Vandenberg, B. "Big Data Analysis for Sustainable Agriculture on a Geospatial Cloud Framework". *Front. Sustain. Food Syst.* vol. 3, art. 54. pp.1–13, July 2019.doi: 10.3389/fsufs. 2019.00054
33. Khan, N., Ray, R.L., Sargani, G.R., Ihtisham, M., Khayyam, M., Ismail, S. "Current Progress and Future Prospects of Agriculture Technology: Gateway to Sustainable Agriculture." *Sustainability*, vol. 13, 4883. pp. 1–13, 2021. https://doi.org/10.3390/su13094883

Chapter 11

# Green IoT for smart agricultural monitoring

*Ushveen Kaur*

## 11.1 IMPORTANCE OF IoT

The Internet of Things (IoT) is made possible by hardware, goods, and devices that have electronics, sensors, firmware, and software built in. This allows the IoT to store, gather, and send data to other platforms or devices over the internet. Through device connectivity and information sharing capabilities, the IoT enhances productivity, efficiency, and adaptability. IoT devices have the potential to screen patients remotely in the medical field, providing medical professionals with timely and essential information. In smart cities, IoT technologies contribute to improved infrastructure, traffic management, and energy efficiency. Additionally, in agriculture, IoT devices aid in precision farming by optimizing resource usage. The IoT's significance lies in its ability to foster innovation, streamline processes, and create a connected ecosystem that enhances decision-making and overall quality of life.

The IoT stands as a ground-breaking technology, propelling advancements in telecommunications and enhancing the quality of life for people across the globe. Furthermore, the IoT holds the promise of making a substantial contribution to the global economy. Projections indicate that the economic influence of the IoT on a global scale could range from USD 2.7 trillion to 6.2 trillion by the year 2025. Experts suggest that the IoT holds the capability to transform various smart domains, encompassing modern healthcare applications, intelligent homes, smart cities, smart agriculture – advanced agricultural practices.

## 11.2 GREEN IoT

With the expansion of IoT, telecom operators face the challenge of facilitating its growth with minimal hurdles. Presently, the number of internet-enabled objects surpasses the global population, and this figure is expected to rise in the near future [1]. By 2026, it is anticipated that machine-to-machine

DOI: 10.1201/9781003628392-11

(M2M) communication will account for 48% of all internet traffic. This is because of the Web of Things' capacity to permit genuine items to speak with one another and complete exercises all alone. Actuators and savvy sensors like Radio Frequency Identification (RFID) are vital for this cycle since they use detecting innovations to send information that they have accumulated from their current circumstances to the web. The coordination of decisions and the information received or transmitted through these sensors allows networked objects in the IoT to execute tasks.

We know each active radio-frequency device used requires a small amount of electrical power for its operation. However, when billions of these network devices are interconnected to make our IoT network, they consume energy daily, and a substantial volume of sensor-transmitted data points needs processing in large data centres. Significant processing and analytics capabilities become essential, potentially leading to high energy consumption. Consequently, the energy consumption of large-scale IoT systems needs careful consideration. To address these energy concerns, the concept of the "Green Internet of Things" (GIoT) has emerged, aiming to enhance energy efficiency and reduce carbon dioxide ($CO_2$)emissions associated with IoT technology.

Due to the IoTs' potential effects on the economy and environment in the coming years, new methods and approaches will be required to meet the energy needs of billions of devices. For the two specialists and providers, the importance of the GIoT has expanded as conventional energy sources run out and energy requests develops. GIoT aims to boost energy efficiency across all aspects of IoT systems, from design to implementation [2]. This has spurred the need to develop various methods for enhancing energy efficiency in IoT, resulting in a broad spectrum of challenges and mitigation techniques. The areas of concern in GIoT techniques can be broadly categorized into two main groups [3]: (i) software, focusing on reducing energy waste through better resource utilization and (ii) hardware, with the goal of improving energy efficiency in smart devices.

## 11.3 WHAT ARE THE TECHNIQUES USED IN GIoT?

We will discuss the energy optimization techniques used in GIoT under five broad IoT implementation categories, namely:

a. Wireless sensors
b. M2M communication
c. RFID
d. Integrated circuits (IC)
e. Microcontroller unit (MCU)

In the following section, we discuss the aforementioned categories in detail.

## 11.4 WIRELESS SENSORS

Wireless sensors are battery-operated, minimal expense gadgets that might convey customary information about their current circumstances or identify explicit occasions. They are a vital piece of IoT innovation. Wireless sensors are fundamental for independently sending information over significant distances consistently, particularly for crucial applications like installed frameworks, traffic announcing, weather condition prediction, and water quality observing. These sensors are powered by batteries, but maintaining or changing the batteries can be costly and time-consuming, especially in remote or difficult-to-reach locations [4]. In specific conditions, geological constraints make battery substitution troublesome. Therefore, in order to ensure the continued viability of wireless sensors, it is absolutely necessary to develop technologies that use less energy in order to extend the life of batteries and cut down on the price of their replacement.

Techniques that are used to implement GIoT wireless sensors are:

a. **Optimizing Radio Energy Consumption:** The primary cause of battery depletion in sensor nodes is the radio unit. The radio unit's consumption of energy is caused by two things:
   i. the amount of energy required to run the circuit and
   ii. the amount of energy required to transmit the signal.

   It is a well-known fact that the power required for the circuit itself matters more than the power needed to transmit the signal across shorter distances. Longer transmission lengths, however, cause the transmitted signal's power to become the main determinant of energy usage. A number of research studies have investigated ways to improve energy efficiency through dynamic transmission power level adjustments [5, 6]. The authors of a study by [7] suggested a power-saving cooperating architecture in which nodes with more energy left over can boost their transmission power, while other nodes can lower their transmission power. This method can improve connections and lessen interference, but it can cause a delay since it might take more hops to forward a message. Collaborative communications between adjacent sensor nodes can mitigate latency concerns by establishing a simulated multiple-antenna setting, providing geographical variety, and enabling data retransmission.
b. **Turning on/off technique on the basis of sleep mode:** In the field of communication and information technology, turning on the sleep mode or turning off idle wireless services and equipment has become a common method of reducing power usage. This method works well for conserving a significant quantity of energy. By utilizing the technique of turning nodes on and off, the suggested solution makes use of the large number of sensor nodes that are deployed within a constrained coverage area with the goal of improving sensor nodes'

energy efficiency and extending battery life. When a system exhibits a significant disparity between its sleep power and active power, substantial energy savings can be achieved. However, there are scenarios where the active power may surpass the sleep power, either due to high power per event or frequent occurrences of active power events. However, coverage is an important factor in this strategy to guarantee sufficiency with the other nodes that are still in operation. Khan, M.A. et al. proposed a solution that activates just a portion of the nodes that have the least amount of duplication in their coverage regions in an effort to reduce the consumption of network energy while maintaining coverage.

c. **Turning on/off technique on the basis of Duty Cycle**: The following categories apply to the duty cycle technique, which involves periodically turning on and off a sensor node based on network activity: (i) a scheduled meeting and (ii) on demand. It is important to keep in mind, though, that although a low-duty cycle might save a significant amount of energy, it may also cause prolonged delays in communication. Protocol parameters can be pre-configured for simplicity prior to deployment in order to minimize delays, but this comes at the expense of less flexibility. As an alternative, these settings can be dynamically adjusted to conform to the traffic situation at hand. Additionally, as explained in [9], the active duration of nodes may depend on a number of variables for the best power usage, including the volume of traffic, buffer overruns, delay needs, or collected energy.
d. **Using Renewable Sources of Energy**: Technological developments in renewable energy are critical to realizing the fundamental characteristics of wireless sensor sources of power, which include sustainability, dependability, and greenhouse gas emissions reduction. In addition, the integration of renewable energy technology presents a great deal of promise for enhancing the energy utilization of wireless networks of sensors located in remote and agricultural regions. In these settings, battery replacement may be difficult because of physical obstacles like difficult-to-access terrain. With a lifetime of 25 to 30 years, solar cells are known for their great dependability and low maintenance needs. Furthermore, new technologies have made it possible for sensors to capture energy from their environment – such as wind and kinetic energy – and transform it into electrical energy for instant consumption or storage [10]. Depending on the environmental setting in which they are used, renewable energy sources might have rather different properties and capacities. For example, renewable energy sources used outside are different from those used indoors in commercial and industrial environments.

However, it can be difficult to assess an IoT device's capability for energy harvesting without knowing how it will be used and how it operates within the system. Notwithstanding technical considerations, some device categories are commonly considered more or less advantageous for storing energy, albeit cost-effectiveness isn't always a determining factor.

In general, energy prediction techniques are necessary for efficiently controlling available power in renewable energy technologies. However, energy-saving measures may need to be included in addition to renewable energy technologies in order to achieve high dependability. In order to maintain living until the next recharge cycle, sensors need to make estimates about the evolution of energy. As a result, they are able to tune critical factors like the sampling rate and sensor architecture in order to match power consumption to the amount and frequency of renewable energy sources.

e. **Type of Battery used in IoT Devices:** Numerous battery types are available for use in IoT devices, each presenting distinct advantages and disadvantages. When choosing a battery for an IoT device, it is important to take into account certain specifications like operating temperature, dimension, and usage of power. It's also important to consider aspects like price, security, and if replacing or recharging the battery is feasible. Figure 11.1 discusses different types of batteries and their features.
f. **Energy-Saving Cluster Architecture and Routing Protocol:** Utilizing cluster architecture proves to be an effective strategy in enhancing the energy efficiency and scalability of wireless sensors through the

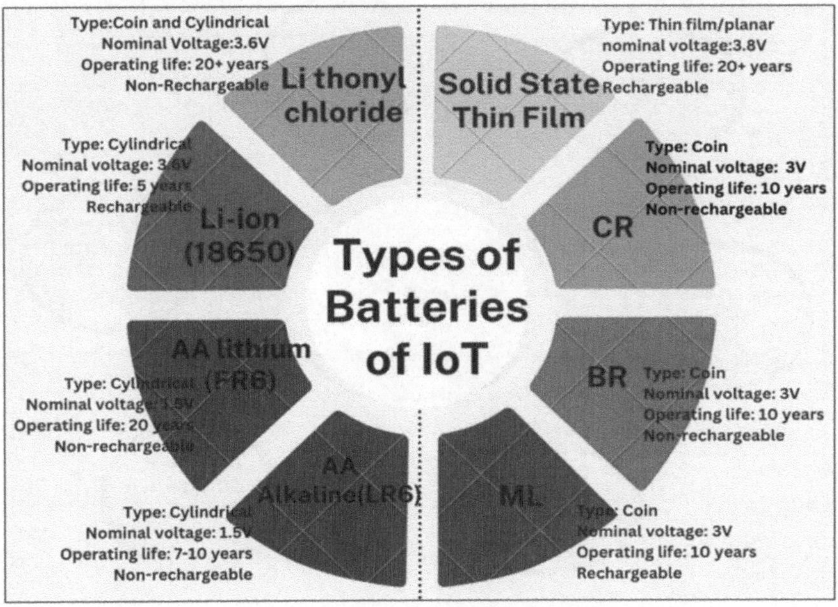


*Figure 11.1* Types of battery.

maintenance of a hierarchical structure. The approach involves grouping sensor nodes into groups, with a selected node known as the cluster leader supervising each cluster. The cluster leader is accountable for supervising the activities of its members and enabling communication among them and other cluster leaders.

Cluster architecture can enhance energy conservation in wireless sensors through following ways:

- Reducing the need for high transmission voltage by shortening communication distances inside the cluster.
- Assigning synchronization and data aggregation duties to the cluster head lowers their energy consumption.
- Assisting in the deactivation of certain cluster nodes while the cluster leader handles forwarding duties.
- Reducing the quantity of transmissions at the cluster head/leader via data fusion.
- Rotating the location of the cluster head allows for an equitable distribution of power consumption among nodes.

We can summarize the wireless sensors energy-saving techniques in Figure 11.2.

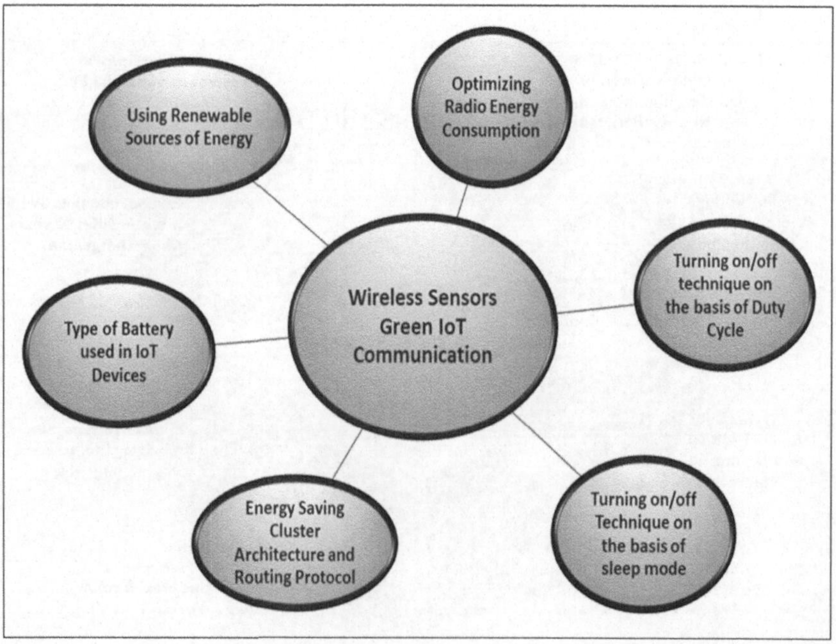


*Figure 11.2* Techniques to implement wireless sensors Green Internet of Things (IoT).

## 11.5 M2M COMMUNICATION

Techniques that are used to implement GIoT in M2M communication are:

**Energy efficiency using data transmission of compressed data** – Compressing the data before sending it can minimize the energy consumption during transmission. Data compression methods that are user-friendly and demand minimal computational resources typically result in lower energy usage [11]. Nevertheless, achieving the right equilibrium between the level of data compression and the associated energy expenditure can be intricate, as techniques emphasizing greater data compression may, at times, necessitate higher energy consumption. Table 11.1 lists the most prevalent data compression methods along with their benefits and drawbacks.

a. **Sharing of resources while data transmission** - M2M devices can cooperate in transmitting data and pooling resources like power, bandwidth, and processing capabilities. This agreeable methodology looks to improve framework execution in general and lower energy utilization.

*Table 11.1* Techniques used for data compression.

| Data Compression Technique | Advantages | Disadvantages |
|---|---|---|
| Huffman coding | • Lossless compression<br>• Simple to implement<br>• Good compression ratios | • Slow speed<br>• Not suitable for real-time data<br>• Requires additional space to prefix data |
| Run-length encoding (RLE) | • Computationally less complex<br>• Works for lossless and lossy compression<br>• Low memory requirements | • Compression ration not good<br>• Works well for *only* repetitive data |
| Delta encoding | • Computationally less complex<br>• Works for lossless and lossy compression | • Not suitable for sensitive data<br>• May lose original information |
| Predictive coding | • Computationally less complex<br>• High compression ratio for related data | • Not suitable for sensitive data<br>• May lose original information |
| Transform coding | • High compression ratio for images, audio<br>• Works for lossless and lossy compression | • Computationally complex<br>• Not suitable for real-time streaming data |

b. **Using machine learning (ML) algorithms on data being transmitted** - ML techniques can be used to improve data transmission speeds and reduce data redundancy in multicell networks. The strategy utilized will depend on the particular necessities of the M2M application, considering factors like the volume and intricacy of the accessible information as well as the necessary level of precision. To find the best calculation for a particular M2M application, it is important to test a few ones and assess how well they work.
c. **Scheduling of M2M devices in power-efficient mode by minimizing energy usage of the device** - In M2M communications, energy-effective data transmission is achieved by effective power scheduling. It is feasible to extend the battery life of M2M devices and improve overall performance and reliability by carefully controlling their energy usage.

   Optimizing M2M device energy consumption while fulfilling performance targets can lead to power-efficient scheduling in M2M communications. These algorithms consider variables including the amount of battery life left, the importance of the data, and the transmission technology's energy efficiency [12].
d. **Scheduling of M2M devices in power-efficient mode by using power-saving modes** – This method uses M2M devices' power-saving settings when there is little activity. These modes include idle modes, in which the device lowers its working frequency to save energy, and sleep modes, in which non-essential components are turned off to preserve power.
e. **Task/Computation Offloading** – Transferring computationally expensive tasks from M2M devices to outside resources with the goal of reducing energy consumption is known as task or computation offloading. In M2M communications, computation offloading provides a number of benefits for energy-efficient data transmission. One notable benefit is the reduction in the M2M device's energy consumption by decreasing the workload on its processor [13]. Additionally, computation offloading enables M2M devices to execute tasks beyond their individual capabilities, overcoming limitations in resources like processing capability or memory [14]. This capability allows M2M devices to do more complex tasks, such as data and prediction analysis, which improves the M2M system's overall performance.

   There are multiple ways to implement task/computation offloading; a few of them are:
   i. **Computation/Task offloading based on distance** – A distributed computing architecture that enables more effective and resource-efficient use by locating processing and data storage closer to the devices and people that need them.
   ii. **Computation/Task offloading based on connecting network edge** – Tasks requiring a significant amount of processing can be offloaded

to a neighbouring edge device, such as a cloud node linked through a network edge or a gateway connected directly, using a local area network (LAN) or Wi-Fi network.

iii. **Computation/Task offloading based on cloud connection** – This category of decreasing the computation can be done over a larger network too, like a Wide Area Network (WAN) or over the complete internet. In this offloading, tasks demanding significant computation are sent to a remote server or a cloud computing machine or a server.

Figure 11.3 shows the types of task/computation offloading.

f. **Using energy saving/efficient hardware** – Efficient hardware plays a crucial role in M2M communication, particularly in the GIoT domain, where devices often need to operate for extended periods on limited power sources like small batteries. Various hardware components are employed to enhance the energy efficiency or enhance the energy-saving capabilities of environmentally friendly green M2M networks.

g. **Power control while data transmission** – Minimizing the energy required for the transmission of data becomes possible by carefully adjusting the transmission voltage current of the M2M devices. A multitude of strategies, including adaptive modulation, duty cycling, routing taking into account power requirements, and adaptive coding, may be utilized to control power and improve the energy consumption of M2M transmission of data.

We can summarize the M2M energy-saving techniques in Figure 11.4.

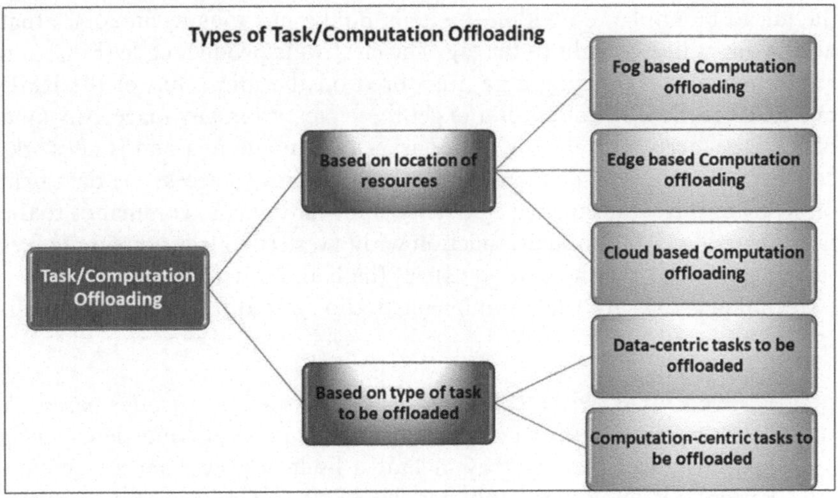


*Figure 11.3* Types of task/computation offloading.

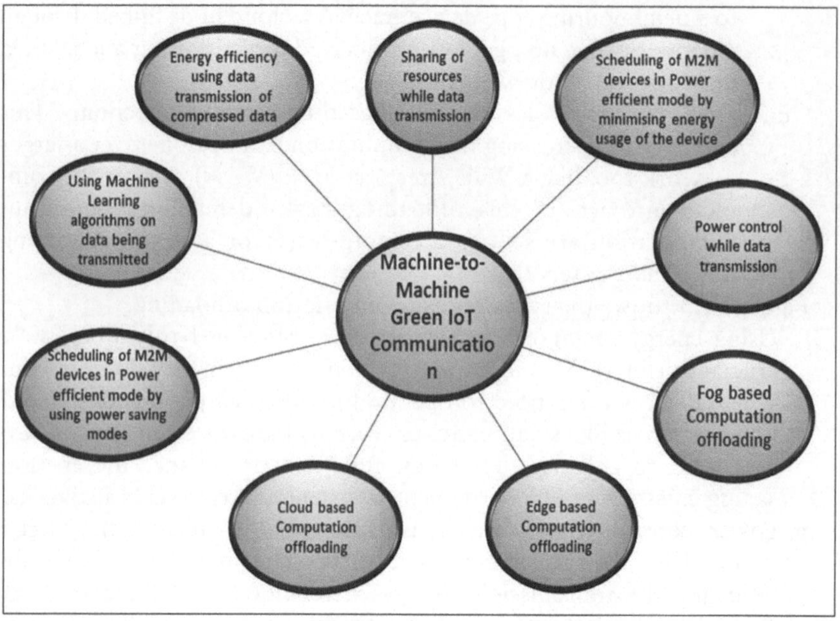


*Figure 11.4* Techniques to implement machine-to-machine (M2M) Green Internet of Things (IoT).

## 11.6 RFID

Different RFID tags, reader equipment, and middleware interaction make up an RFID setup. Things that need to be tracked are fitted with RFID tags, which are made up of an antenna and a microchip. RFID scanners allow the tag to be implanted within the item, unlike previous technologies that need a direct line of sight to the tag. The electronic product code (EPC) and any other crucial tracking data are stored on the microchip of the RFID tag. RFID readers can monitor and get these facts from any place. Any kind of RFID reader – portable or stationary – has an antenna and is network-connected. It acts as a conduit for business systems to access the data held on RFID tags by transmitting electricity, information, and commands to the tags. Generally, RFID systems function within restricted transmission ranges (few meters) and over many frequencies (high and low frequency bands).

Techniques that are used to implement GIoT radio frequency communication are -

a. **Passive RFID Device:** RFID tags may be divided into groups based on how they receive their power: Active RFID tags can send signals over long distances because they include a built-in power source, often a battery. On the other hand, passive RFID tags rely on power generated from the RFID reader's signal to operate its microprocessor as they don't have an internal power source. Passive tags therefore have a limited transmission area coverage. An RFID tag that is semi-passive has an energy source, but it is only used to supply power to the microchip and not to transmit signals [15].

RFID passive systems do not require an internal power source since they use tags that draw power indirectly from the reader's electromagnetic signal. Due to their cost-effectiveness, with a lower cost per tag, passive RFID tags are a preferred solution across various industries and agriculture.

Because of their low consumption of electricity, passive RFID tags are especially well suited for wireless applications. Still, their working range is limited; the most sophisticated passive tags can only go 8–15 meters. Furthermore, one of the main factors limiting RFID transmission range in passive tags is path loss, which is an important aspect of wireless communication.

RFID readers are the main component of the network as they act as passive RFID wake-up radios. The wake-up radios working as passive RFID are triggered to awaken the node by the energy generated by the reader transmitter. But because RFID readers use a lot of power, it is not feasible to use them for every sensor; instead, they are useful only in single-hop scenarios.

b. **Active RFID Device:** In an active RFID system, battery-powered RFID tags are employed to consistently transmit their own signals. These battery-powered tags are utilized for precise real-time tracking of assets, especially in high-speed environments like tolling. Although more costly than passive tags, they offer a significantly larger scanning range.

   The energy storage system is a critical consideration in the context of an active RFID system, with batteries and capacitors being the two most prevalent options.

c. **Use of batteries for energy storage:** Since their creation, batteries have been a common source of power for a wide range of products and are still in widespread use today. But even with rechargeable batteries, which eventually lose their ability to hold energy, they have a certain lifespan and must be replaced after a given amount of time. Another important factor to take into account when choosing a storage device is its size and form. However, form, size, and capacity are frequently trade-offs because bigger capacities usually translate into larger device sizes. This antagonistic connection might not meet the needs of RFID contexts, where a small, flexible gadget is necessary. The goal of research has been to create small energy storage systems with improved flexibility and no capacity loss. The development of batteries with the necessary form factor (size and shape) for RFID applications has been made easier by the introduction of thin-film technology. With the use of this technology, exceptionally thin electronic devices that are ideal for thermo-electric micro-systems may be produced. Research suggests that thin-film batteries are especially well suited for uses where there are temperature variations, such those that occur close to the human body [16]. In conclusion, thin-film technique is flexible and appears to be a good option for RFID tags.

d. **Use of capacitors for energy storage:** Capacitors stand out as highly energy-efficient storage devices, maintaining their capacity over an extended period, which renders them well suited for specific applications. Nevertheless, the size of a capacitor significantly influences its capacity, leading to a trade-off where capacitors often prioritize longer lifespan over form factor considerations.

   Furthermore, because of how easily they may be affected by variations in voltage and current, capacitors can be rather unreliable. The advent of ultra-capacitors and super-capacitors, which combine the longevity of conventional capacitors with the dependability as well as efficiency of batteries that are rechargeable, is a recent development in capacitor storage technology that offers the capacity to store significant amounts of energy. Figure 11.5 discusses differences between capacitors and batteries with respect to capacity, energy density, reliability, and form factor. Designers opt for these ultra-capacitors due to their high energy densities, surpassing those of conventional capacitors by several orders of magnitude.

   Hybrid systems that integrate various storage devices have also been explored, combining the strengths of different components. Such models have undergone scrutiny and experimentation yielding promising outcomes.

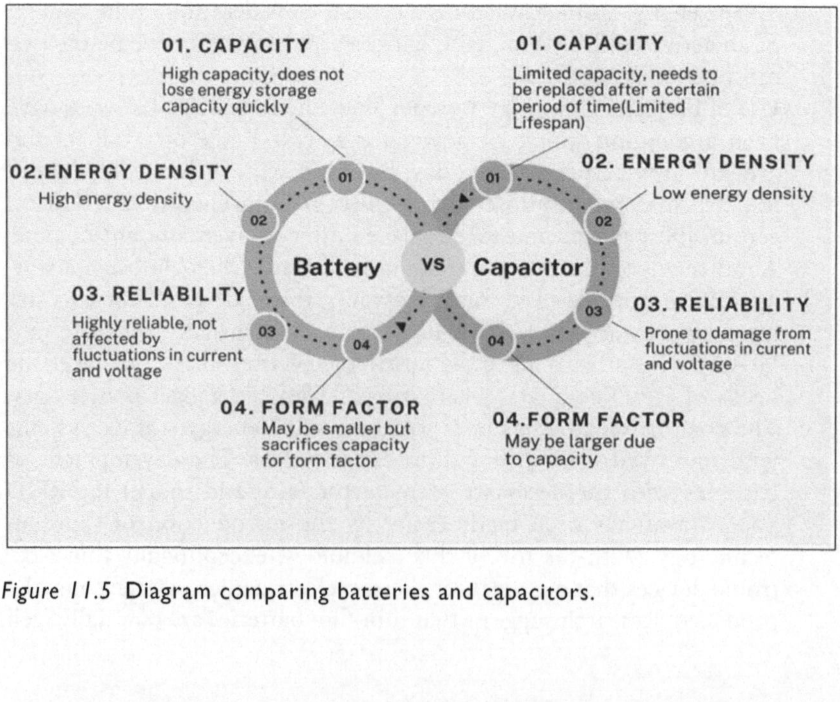


*Figure 11.5* Diagram comparing batteries and capacitors.

## 11.7 SMART AGRICULTURE

Agriculture stands as the primary contributor to India's economy, constituting 18% of the gross domestic product and engaging approximately 57% of the rural population. Despite an increase in India's overall agricultural output, the proportion of individuals involved in farming decreased from 71.9% in 1951 to 45.1% in 2011 [17]. The Economic Survey of 2018 projects that by 2050, the share of workers in agriculture will fall to 25.7% of the overall employment. Due to obstacles including growing cultivation expenses, low per capita production, inadequate soil upkeep, and a trend towards non-farming or more profitable employment, agricultural families in rural regions are increasingly losing their next generation of farmers. The world is about to experience a digital revolution; therefore, now is the perfect time to introduce and enable digital connection for farmers by integrating wireless technology into agricultural landscapes.

Each crop field possesses distinct critical characteristics, including soil type, irrigation flow, nutrient presence, and pest resistance. These factors are individually assessed in terms of both quality and quantity for a specific crop. To optimize crop production within the same field, it is essential to consider both spatial and temporal variations, incorporating strategies such as crop rotation and aligning with the annual crop growth development cycle.

Figure 11.6 shows main features of agriculture where technology can play a role.

Precise monitoring enabled by the strategic placement of sensors may turn farming as a whole into an intelligent and economic activity. Data is often gathered by sensor-equipped drones, robotic weed eaters, and driverless harvesting equipment. But because agriculture is so widespread, technological

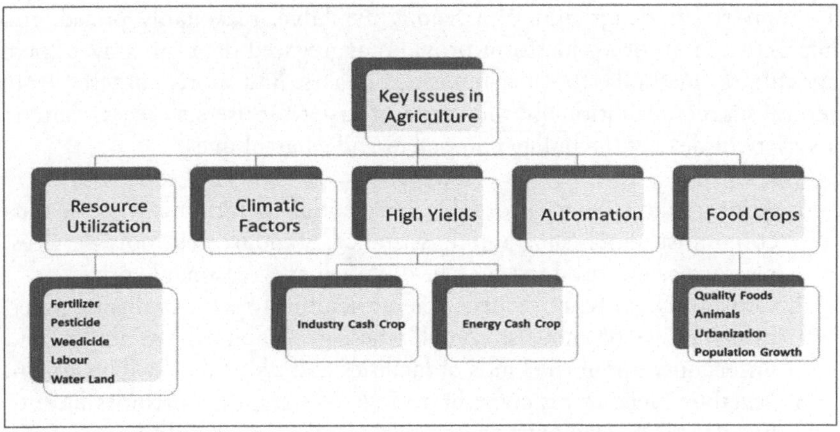


*Figure 11.6* Key issues in agriculture where technology can help.

advancements are necessary to guarantee sustainability while minimizing its negative effects on the environment. Farmers may evaluate the various demands and requirements of crops from a distance using wireless communication-enabled sensor technology, which enables them to make educated decisions without having to physically be in the fields [18].

The agricultural industry of today uses modern technology including cloud computing, artificial intelligence (AI), big data analysis, IoT, and remote sensing. By creating affordable sensor and network platforms, the integration of these new technologies has greatly improved agricultural operations. The main objective is to minimize the impact on the environment while maximizing production efficiency and using less energy and water [19]. Big data in smart farming helps farmers make educated decisions by giving them deep, real-time insights on agricultural circumstances [20]. Farmers may make more educated decisions with the use of real-time programming that is integrated into IoT devices and imbued with AI principles [21].

Smart farming fosters precision agriculture through advanced technology, empowering farmers to monitor plants remotely. Because this method increases the productivity of the farming labour through the automated use of sensors and gear, it improves agricultural operations including crop yields and harvesting [22]. Agriculture is experiencing a technological revolution as a result of the deployment of these technologies, which convert manual agricultural techniques into automated operations. The current technological landscape has reshaped farming practices, with conventional techniques giving way to innovations introduced by the IoT [23].

## 11.8 PRINCIPLES OF IOT UTILIZATION IN AGRICULTURE

To support smart agriculture, a readily available, reasonably priced, and interactive monitoring platform provides aggregated data on conventional agricultural methods, tools, crop pests, diseases, and more, gathered from several sources. Additionally, this makes it easier for users to access data on a variety of devices, including computers and smartphones.

1. Affordability: Being affordable is essential to farming success; thus costs must match substantial assistance. Standardized tools, services, platforms, and products may be obtained at a reasonable price.
2. Constantly evolving models: The agricultural sector is distinguished by its range of models, complexities, geographical variability, and uncertainty about the kinds of facilities and crops that will be grown.
3. Scalable: Since farms come in a range of sizes, the outcomes have to be adaptable. Deployment and testing planning should be gradually increased at the lowest possible cost.

4. Sustainable: With the challenges of the global economy and competitiveness, sustainability is an important concern.

## 11.9 TECHNOLOGY THAT IS USED IN SMART FARMING

a. **Global Positioning System (GPS):** GPS receivers are able to determine their current location and provide continuous locations when moving since GPS satellites send out signals. Farmers are able to detect particular field data, such as insect occurrences, soil kinds, weeds, and other challenges, with the use of this precise location information. The technology makes it easier to identify different fields, which enables the accurate administration of inputs like seeds, fertilizers, pesticides, herbicides, and water to certain fields.
b. **Sensor Technologies:** The assessment of soil texture and structure, amounts of nutrients, vegetation, moisture, vapor, the environment, temperature, and other factors is done through the use of a variety of methods, such as photoelectricity, electromagnetic radiation, electrical conductivity, and ultrasound waves. Monitoring water levels, classifying pests and weeds, detecting stress in soil and plant conditions, and differentiating between crop varieties all depend heavily on data from remote sensing. Sensing technologies frequently use sensors for moisture, temperature, soil condition monitoring, the flow of air, location, $CO_2$ pressure, light, and dampness. The appropriateness of sensors for agricultural applications is influenced by several key qualities, such as computational efficiency, durability, movement, retention, and dependability. Depending on the application, a variety of sensors, including acoustic, airflow, electrochemical, electromagnetic, mechanical, and optical sensors, are utilized.
c. **Soil Sampling:** In soil sampling, a field is divided into tiny regions or cells using a grid overlay, and samples are taken from designated parts. In order to represent the full zone, composite samples are gathered from each considerably smaller region at the grid line intersections. In order to estimate values at non-measured sites between sampled points, interpolation techniques are used to map soil-test results acquired from grid sampling. Because each field has different levels of potassium and phosphorus, different fertilization techniques are needed to optimize nutrient management by applying fertilizers and manure uniformly and enhancing precision agriculture.
d. **Geographic Information System:** Field characteristics including crop output, surfaces and underground drainage, terrain, irrigation, soil types, and nutrient levels are all covered in depth by GIS databases. It also creates links between factors that affect a crop in a particular farming area. In addition to its use for storing and presentation, GIS

is used to combine and change data layers in order to make decision-making easier. This allows for the evaluation of both current and alternative management techniques.

e. **Crop Management:** Satellite photos provide information on how the topography of a field affects crop performance and alterations in soil conditions. As a result, farmers are able to closely monitor production elements that lead to higher output and efficiency, such as seeds, fertilizers, and insecticides.
f. **Plant and Soil Sensors:** A key element of precision agriculture is sensor technology, which provides data on the fertility, water condition, and characteristics of the soil. As a result, new sensors that differ from those that are already on the market have been developed with desirable qualities. Plant wearables and soil sensors monitor physical and chemical signals in the soil in real time, such as temperature, humidity, pH, and contaminants. They provide information to fight biotic and abiotic challenges, improve agricultural yields, and optimize crop growth circumstances.
g. **Precision Irrigation:** Irrigation systems have advanced recently with the addition of specialized gear with wireless connection, sensor technologies, GPS-based controls, and motion control. These systems are designed to measure irrigation factors including flow and pressure, as well as soil and climate conditions. The objective is to increase crops' water use efficiency. Even though these technologies show a lot of promise, more work needs to be done before they can be extensively used in commercial applications.

## 11.10 BASIC FRAMEWORK OF IoT-BASED AGRICULTURE

A plentiful, secure supply of agricultural goods with low energy usage is the aim of sustainable agriculture. Monitoring the growth cycle as well as the full life cycle of agricultural goods is crucial to achieving this. Practically speaking, a lot of problems with the quality and safety of agricultural goods are caused by outside variables, such as distribution, storage, transportation, processing, and ingredient supply. Farmers' enthusiasm for smart farming methods may be increased by using IoT systems across the whole life cycle of agricultural goods. This will help farmers analyse the quality of agricultural materials and produce products that are worthy of the market. However, adding more sensors frequently results in higher energy consumption; thus while implementing farm IoT systems, ecological requirements must be taken into account. Applying IoT technologies across the whole agri-products' manufacturing process, upholding green standards, and creating a life cycle framework for GIoT-based farming are therefore imperative.

IoT techniques in life cycle framework of green agriculture:

a. **Availability of Ingredients:** The implementation of IoT methods and processes in ingredient supply has the potential to enhance the efficiency and efficacy of ingredient management. Farmers employing IoT in their ingredient inventory can monitor the levels and conditions of various ingredients in real time, enabling precise and automatic ordering based on inventory control policies. When supplies like fertilizers and seeds arrive, their quality and amount may be quickly ascertained by evaluating ingredient tracing systems and using RFID smart tags. Additionally, IoT systems offer benefits to ingredient suppliers by facilitating the cultivation of seeds and the production of fertilizers.
b. **Managing and Arranging:** To increase farmers' profits, effective agri-product categorization marketing must be put into place. Conventional categorization techniques are labour-intensive and inefficient. Non-destructive determination methods, such as scent sensors and near-infrared spectroscopy analysis, have been used in agriculture IoT systems for autonomous fruit categorization. Furthermore, for perishable agricultural goods, the processing environment needs to be closely observed and managed. In this case, improving processing for various fresh produce kinds that demand distinct processing settings is an application. Using information from ingredient supply and growth linkages to create intelligent labels, which are then attached to packaged agricultural goods, is another IoT strategy in this area.
c. **Delivery and Warehousing:** Agri-product quality fluctuations are greatly influenced by the storage environment. Building IoT-based warehouses is becoming more popular as a way to reduce the loss of fresh agricultural products. Supervisors may track the whereabouts of refrigerated trucks and the condition of agri-products in transit by using GPS and sensor networks installed on their vehicles. When there is a risk to quality from the conditions within refrigerated vehicles, warning signals are generated. In order to facilitate timely receiving plans, downstream retailers and customers may both track their ordered agri-products in real time. Automated classification is still another essential feature that comes from IoT systems, especially for completing purchases for agricultural products online.
d. **Usage:** Agri-products with RFID tags or QR codes on store shelves make it easier for customers to identify the variety and origin, which keeps them from buying shoddy or fake goods. Customers may also obtain complete information from the tracking system about the full life cycle, including records of pesticide and fertilizer use. Supermarket IoT devices help customers quickly identify additives and pesticide residue. Life cycle IoT solutions help the recall procedure become more accurate in case specific agri-products have safety problems. Figure 11.7 expresses the life cycle framework in the most appropriate way.

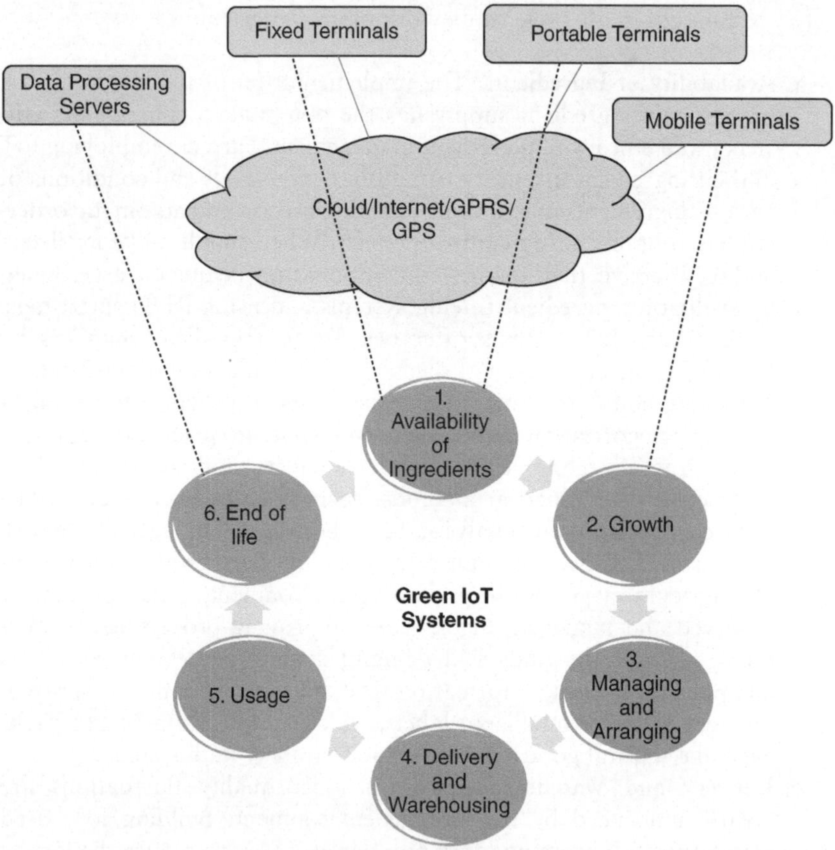


*Figure 11.7* Green Internet of Things (IoT) life cycle.

## 11.11 GIoT TECHNIQUES IN SMART AGRICULTURE

a. **IoT-Controlled Seedling Production System:** IoT-controlled seedling production system is a framework for IoT-based greenhouse seedling production. The framework, its components, and their information transmission throughout the entire system are explained in this section. Information is transmitted via the network using various technologies, such as the internet, GPRS, GPS, and Wi-Fi.

   The system is divided into three sub-functional modules:

1. **Data Collection Module:** The data collecting module's main job is to keep an eye on environmental data both within and outside the greenhouse. Positioned in the greenhouse and on the agriculture meteorological location, IoT sensors continually gather and provide real-time data to the central server, including temperature,

humidity, and illumination intensity. Furthermore, cameras within the greenhouse record the seedlings' current growing state in real time, improving the capacity for remote monitoring and diagnostics.

2. **User Interaction Module:** The administrator and the farmer are the two main users who communicate directly with the IoT system. In addition to maintaining the system, the administrator coordinates production. Among the duties assigned to her are uploading the product catalogue, updating product variety and delivery schedules, assigning demand for seedlings to particular greenhouses based on structure and rotation plans, and planning production activities. The administrator decides on emergency management measures and may visit the site in unique situations, such as large-scale insect infestations or natural catastrophes. When ordering seedlings, farmers or other system users can peruse the product catalogue, place orders, and remotely monitor the environment record and seedling development status in the relevant greenhouse. This feature lessens dispute and builds farmer trust in the conventional seedling production sector.
3. **Data Processing and IoT Equipment Control Module:** The data collecting module or user interaction module needs provides the data that the central server uses to execute statistical and analytical tasks. Users receive a report on the analytical results, or the results are translated into directives for the control cabinet. The control cabinet is where the greenhouse's machinery and outside orders are transferred. The control cabinet receives control commands from the central server, allowing the greenhouse's machinery to operate automatically.

Pre-germination and post-germination are the two processes that follow in the creation of seedlings (open field planting). The entire pre-germination phase takes place in a secluded space called the germination cabinet in order to guarantee a greater survival probability. The crop species determines the set germination period and the consistent conditions within the germination cabinet. After sprouting, the seedling tray is moved to solar greenhouses outfitted with remotely operated machinery and sensors for monitoring the environment in real time.

Standard solar greenhouses are able to reduce temperature variations by isolation from the outside world.

When the climate in the greenhouse isn't conducive to the growth of seedlings, the system uses heaters, windows, and evaporation coolers – ventilate fans and cooling pads placed on the other side of the greenhouse – to regulate the temperature.

Plant growth lights and shade screens are used to enhance the control impact and reduce energy expenses related to extended heater and evaporation cooler operation.

The crop development model and historical data are used to specify the quantity of supplementary and shade light needed over the whole growth cycle in order to handle the interplay between temperature regulation and light. Afterwards, real-time monitoring data is used to guide short-term equipment control.

b. **Livestock Breeding**: IoT techniques exhibit clear advantages in enhancing livestock management for both confined spaces and extensive pastures. One fundamental role of IoT-based farms is to ensure the optimal growth of livestock by monitoring and controlling factors like temperature and humidity. Innovative IoT devices like wearable sensors and three-dimensional accelerometers are facilitating the implementation of emerging directions in this sector, which include healthcare monitoring for animals and behaviour recognition.

By tracking the movements of cattle and keeping an eye on the circumstances that promote herbage growth, IoT devices help to improve pasture management. Additional uses of IoT technologies in livestock breeding include animal identification and feeding log recording.

The IoT is transforming livestock farming by bringing sophisticated tools to assist farmers avoid crop waste and making conventional procedures more cost-effective. Automated feeding, mechanized milking, and increased productivity efficiency via nutrition, animal health monitoring, and instrumentation are examples of recent innovations. Even with these encouraging advancements, there are still big problems. In order to fulfil the increasing demand for animal products, intensive livestock farming is essential; nevertheless, the cramped and constricted space makes it challenging for farmers to precisely identify and monitor animal health. In order to identify health and disease outbreaks early on, comprehend the mechanism of disease transmission, and put preventative measures in place to avoid significant financial losses, it is imperative that these issues be addressed.

c. **Bioscience and hydroponics**: Fish farms that use specialized IoT sensors have found that monitoring water quality indicators, such as pH, salinity, dissolved oxygen, temperature, and level, is easier and more advantageous. Actuators may start the proper processes, such as filling the water and producing oxygen, in regulated fish farms to keep the water quality ideal. Using underwater cameras and ultrasonic detectors makes it easier to identify fish groups and allows for precise feeding. Hydroponics and aquaculture are combined in aquaponics, which is another possible application area for IoT systems. Hydroponics powered by IoT makes it easier to keep an eye on the growing conditions of hydroponic plants as well as aquatic life. Compared to independent aquaculture and greenhouse setups, this frequently calls for more sensors, actuators, and processing rules.

## 11.12 IoT IN BIOSCIENCE/AQUACULTURE

**Water Quality Management:** IoT sensors monitor key parameters such as water temperature, pH, dissolved oxygen, and salinity in real-time, ensuring optimal conditions for aquatic life.

**Automated Feeding Systems:** Smart feeders equipped with IoT technology enable precise and automated feeding, reducing waste and improving feed efficiency.

**Health Monitoring:** Wearable sensors on fish collect data on health and behaviour, allowing early detection of diseases and timely intervention.

**Environmental Control:** IoT devices control factors like lighting, aeration, and water circulation, creating a conducive environment for fish growth.

**Remote Monitoring:** Farmers can remotely manage and monitor aquaculture systems through IoT applications, enhancing operational efficiency.

## 11.13 IoT IN HYDROPONICS/AQUAPONICS

**Integrated Monitoring:** IoT facilitates the integration of aquaponics, where fish farming and plant cultivation coexist in a symbiotic system.

**Environmental Optimization:** Sensors monitor conditions for both aquatic creatures and hydroponic plants, ensuring optimal growth for both components.

**Resource Efficiency:** Automated control systems manage nutrient levels, water circulation, and other variables, optimizing resource utilization.

**Data-Driven Decisions:** IoT-generated data in aquaponics systems can be analysed to make informed decisions, enhancing overall system efficiency.

In summary, IoT in aquaculture and aquaponics enhances precision, efficiency, and sustainability in the management of aquatic ecosystems, ushering in a new era of smart and environmentally conscious farming practices.

## 11.14 CONCLUSION

In summary, the IoT has become a game-changing technology that will have a significant impact on a number of industries, including smart cities, telecoms, agriculture, and the monitoring of the environment. The seamless connectivity and data sharing enabled by IoT devices have significantly enhanced efficiency, productivity, and convenience in numerous applications. In smart cities, IoT technologies have contributed to improved infrastructure, traffic management, and energy efficiency.

However, the rapid proliferation of IoT devices poses challenges, particularly in terms of energy consumption. As the number of internet-enabled objects surpasses the global population, careful consideration of energy needs becomes crucial. Various data compression techniques have been explored to mitigate energy consumption in M2M communications, ranging from collaborative resource sharing to the use of ML algorithms for data optimization. Additionally, power-efficient scheduling, cloud-based computation offloading, and the use of energy-efficient hardware have been identified as effective strategies.

In the context of agriculture, IoT plays a pivotal role in advancing precision farming, monitoring soil conditions, and optimizing crop production. The integration of IoT with sensors, GPS, and remote sensing technologies provides farmers with real-time data for informed decision-making. From seedling production to livestock breeding and aquaponics, IoT applications offer innovative solutions to enhance sustainability and productivity.

As we move forward, addressing the energy challenges associated with the widespread adoption of IoT remains a priority. The development and implementation of advanced techniques and strategies, such as energy-efficient hardware and optimized data compression methods, will be essential for ensuring the long-term success and sustainability of IoT applications across diverse sectors.

## REFERENCES

[1] Vermesan et al., "Internet of things: Strategic research roadmap," 2009; http://www.internet-of-things-research.eu/pdf/IoT_Cluster_Strategic_Research_Agenda_2011.pdf, accessed Apr. 3, 2018.

[2] Chiang, C. T., "A CMOS seawater salinity to digital converter for IoT applications of fish farms," *IEEE T. Circuits-I*, vol. 64, no. 9, 2017, pp. 2591–2597.

[3] Brewster, C. et al., "IoT in Agriculture: Designing a Europe-Wide Large-Scale Pilot," *IEEE Commun. Mag.*, vol. 55, no. 9, 2017, pp. 26–33.

[4] Rault, T.; Bouabdallah, A.; Challal, Y. "Energy efficiency in wireless sensor networks: A top-down survey." *Comput. Netw.* 2014, 67, 104–122.

[5] Lin, S.; Miao, F.; Zhang, J.; Zhou, G.; Gu, L.; He, T.; Stankovic, J.A.; Son, S.; Pappas, G.J. "ATPC: Adaptive transmission power control for wireless sensor networks." *ACM Trans. Sens. Netw.* 2016, 12, 6.

[6] Correia, L.H.; Macedo, D.F.; dos Santos, A.L.; Loureiro, A.A.; Nogueira, J.M.S. "Transmission power control techniques for wireless sensor networks." *Comput. Netw.* 2007, 51, 4765–4779.

[7] Chu, X.; Sethu, H. "Cooperative Topology Control with Adaptation for improved lifetime in wireless sensor networks." *Ad. Hoc. Netw.* 2015, 30, 99–114.

[8] Misra, S.; Kumar, M.P.; Obaidat, M.S. "Connectivity preserving localized coverage algorithm for area monitoring using wireless sensor networks." *Comput. Commun.* 2011, 34, 1484–1496.

[9] de Paz Alberola, R.; Pesch, D. "Duty cycle learning algorithm (DCLA) for IEEE 802.15. 4 beacon-enabled wireless sensor networks." *Ad. Hoc. Netw.* 2012, 10, 664–679.
[10] Sudevalayam, S.; Kulkarni, P. "Energy Harvesting Sensor Nodes: Survey and Implications." *IEEE Commun. Surv. Tutor.* 2010, 13, 443–461.
[11] Azar, J.; Makhoul, A.; Barhamgi, M.; Couturier, R. "An energy efficient IoT data compression approach for edge machine learning". *Futur. Gener. Comput. Syst.* 2019, 96, 168–175.
[12] El-Hosseini, M.; ZainEldin, H.; Arafat, H.; Badawy, M. "A fire detection model based on power-aware scheduling for IoT-sensors in smart cities with partial coverage." *J. Ambient. Intell. Humaniz. Comput.* 2021, 12, 2629–2648
[13] Atzori, L.; Bellido, J.; Bolla, R.; Genovese, G.; Iera, A.; Jara, A.; Lombardo, C.; Morabito, G. "SDN & NFV contribution to IoT objects virtualization." *Comput. Netw.* 2018, 149, 200–212.
[14] Aucinas, A.; Crowcroft, J.; Hui, P. "Energy efficient mobile M2M communications." In *Proceedings of the 4th Extreme Conference on Communication*, Zurich, Switzerland, 10–14 March 2012; Volume 12.
[15] Khan, M.A.; Sharma, M.; Prabhu, B.R. "A survey of RFID tags." *Int. J. Recent Trends Eng.* 2009, 1, 68.
[16] Yeo, J.; Moon, S.-G.; Jung, J.-Y. "Antennas for a battery-assisted RFID tag with thin and flexible film batteries." *Microw. Opt. Technol. Lett.* 2007, 50, 494–498.
[17] Reddy, T.; Dutta, M. "Impact of Agricultural Inputs on Agricultural GDP in Indian Economy." *Theor. Econ. Lett.* 2018, 8, 1840–1853.
[18] Lin, J.; Yu, W.; Zhang, N.; Yang, X.; Zhang, H.; Zhao, W. "A Survey on Internet of Things: Architecture, Enabling Technologies, Security and Privacy, and Applications." *IEEE Internet Things J.* 2017, 4, 1125–1142.
[19] Ferrandez-Pastor, F.J.; Garcia-Chamizo, J.M.; Nieto-Hidalgo, M.; Mora-Pascual, J.; MoraMartinez, J. "Developing ubiquitous sensor network platform using Internet of Things: Application in precision agriculture." *Sensors* 2016, 16, 1141.
[20] Wolfert, S.; Ge, L.; Verdouw, C.; Bogaardt, M.J. "Big data in smart farming—A review." *Agric. Syst.* 2017, 153, 69–80.
[21] Liakos, K.G.; Busato, P.; Moshou, D.; Pearson, S.; Bochtis, D. "Machine learning in agriculture: A review." *Sensors* 2018, 18, 2674.
[22] O'Grady, M.J.; O Hare, "G.M.P. Modelling the smart farm." *Inf. Process. Agric.* 2017, 4, 179–187.
[23] Quy, V.K.; Hau, N.V.; Anh, D.V.; Quy, N.M.; Ban, N.T.; Lanza, S.; Randazzo, G.; Muzirafuti, A. "IoT-Enabled Smart Agriculture: Architecture, Applications, and Challenges." *Appl. Sci.* 2022, 12, 3396.

Chapter 12

# Agricultural farming decision support system using artificial intelligence

## A comparative analysis

*Amaresh Sahu, Biswaranjan Acharya and Partha Sarathi Sahoo*

## 12.1 INTRODUCTION

Crop yield prediction, as described in this chapter, is indeed a crucial application of machine learning (ML) and data analysis in the agriculture industry. The importance of Crop Yield Prediction (CYP) [1] lies in its crucial role in the agricultural market, particularly in a rapidly changing environment. Absolutely, accurate crop yield predictions are invaluable to farmers and agricultural stakeholders. Let's delve deeper into how these predictions benefit agriculture like informed decision-making that gives the accurate crop yield predictions that empower farmers to make data-driven decisions at every stage of the crop lifecycle, from planting to harvesting and post-harvest management. Global optimization of planting decisions helps with crop selection by enabling farmers to choose the most suitable crops for a given season and location based on predicted yields. Fertilizer application predictions [2] can guide farmers in applying fertilizers and soil amendments precisely, preventing overuse and environmental harm. Pest and disease management predicts accurate forecasts that can help farmers prepare for potential pest and disease outbreaks, reducing the need for chemical interventions. Harvest planning means farmers can plan their harvests based on predicted yields and maturity, optimizing labor and machinery usage. Sustainability involves minimizing environmental impact by optimizing resource use. Crop yield predictions contribute to more sustainable farming practices by reducing resource waste and environmental harm.

Market basket analysis and forecasting can also be applied to market trend predictions and decision-making. Farmers can analyze market trends, demand patterns, and price fluctuations to make informed decisions about crop selection and timing for harvest and sale. Climate change mitigation in agriculture is both impacted by and contributes to climate change. Decision trees are versatile machine learning (ML) algorithms used in various applications within smart farming. Here are some specific applications of decision trees in the context of agriculture and smart farming: In species identification decision trees can be trained to identify different plant and

 DOI: 10.1201/9781003628392-12

animal species based on specific features such as leaf shape, size, or color. This is useful in monitoring biodiversity in agricultural ecosystems. In precision agriculture decision trees help in optimizing the use of resources in precision agriculture. By analyzing various factors such as soil quality, weather conditions, and crop types, decision trees can assist in making decisions related to irrigation, fertilizer and pesticide application, maximizing yield and minimizing resource usage. In crop health monitoring, decision trees can process data from Internet of Things (IoT) devices and remote sensors to monitor the health of crops. By analyzing parameters like soil moisture, temperature, and humidity, these algorithms can alert farmers to any deviations from the optimal conditions, enabling proactive measures to maintain crop health. In yield prediction decision trees predict crop yields based on historical data, weather patterns, soil quality, and other variables. Accurate yield predictions help farmers plan harvest and post-harvest activities, storage, and distribution effectively. In quality grading also decision trees are used to grade and sort agricultural produce based on quality attributes. For instance, fruits and vegetables can be graded based on size, color, and blemishes, helping in efficient packaging and marketing strategies. In optimal harvesting time series prediction decision trees can analyze data related to crop maturity, weather forecasts, and market demands to suggest the optimal time for harvesting. This ensures that crops are harvested at the peak of their quality and market value. In livestock management decision trees are used in livestock farming to monitor animal health and behavior. By analyzing data from sensors attached to animals, these algorithms can detect signs of illness or distress, allowing farmers to provide timely veterinary care. In weed detection and management, the decision trees can process images of fields to identify and differentiate between crops and weeds. This information is valuable for implementing targeted weed control measures, reducing the need for excessive herbicide usage.

Support vector machine (SVM) [3] and support vector regression (SVR) [4] are supervised ML algorithms used for classification and regression tasks, respectively. They work by finding the optimal hyperplane (in the case of SVM for classification) or a regression line (in the case of SVR) that best divides or approximates the data into different classes or predicts numerical values. This provides hyperplane selection where SVM aims to find a hyperplane that best separates the data points of different classes. This hyperplane is chosen in a way that maximizes the margin, i.e., the distance between the hyperplane and the nearest data points from either class, also known as support vectors. If the data is not linearly separable, SVM uses a kernel trick, which maps the data into a higher-dimensional space where a separating hyperplane might exist. Kernel trick in SVM can use different kernel functions (such as linear, polynomial, radial basis function—RBF, or sigmoid) to transform the input features into a higher-dimensional space. This transformation can make the data linearly separable in that higher-dimensional space. SVM is primarily used for binary classification problems but can be extended to handle multiple classes using techniques like one-vs-one or one-vs-rest.

Artificial neural networks (ANN) [5] and deep learning (DL) [6] have various applications in image processing and analysis. ANN is used in tasks such as weed recognition and plant disease detection. DL leverages advanced models, such as convolutional neural networks (CNNs), in ensemble learning environments for image-based tasks. It emphasizes the importance of using a robust heterogeneous ensemble learning environment (HELE) [7] that combines various effective ML algorithms to improve prediction accuracy and reliability. Here's a breakdown of the key points and objectives from the proposal. It utilizes IoT sensor networks to collect data on various agricultural parameters such as soil moisture, temperature, humidity, etc., and integrate this data with advanced machine learning (AML) techniques for analysis and prediction. This ensemble approach aims to leverage the strengths of different ML supervised algorithms to enhance prediction accuracy. The excerpt we provided outlines a research proposal focusing on the integration of IoT technologies and AML algorithms for smart farming and agricultural decision-making. The chapter highlights the various aspects of digital agriculture and the impact of digital technologies on farming practices. Here's a summary of the key points mentioned. The Internet of Things (IoT) plays a crucial role, utilizing sensors—including food and soil sensors—along with global positioning system (GPS), global navigation satellite system (GNSS), radio frequency identification (RFID), and other IoT technologies to enable precise and localized farming practices. Precision agriculture technologies like variable-rate application (VRA) enable farmers to apply inputs such as water, fertilizer, pesticides, etc., precisely to localized regions of their farms, improving efficiency and reducing costs. Studies indicate that precision agriculture can lead to increased yields. Equipment sharing platform, which is a digital platform, facilitates the sharing of expensive farming machinery, allowing farmers to access equipment without major investments. This approach optimizes the use of resources and increases productivity.

## 12.2 RELATED WORK

An ANN [5] algorithm was applied, and additional deep learning (DL) algorithms were explored, including convolutional neural networks (CNN), long short-term memory (LSTM), and deep neural networks (DNN). We found that LSTM, DNN, and CNN are highly preferred DL techniques for their application. It's clear from the information provided that both studies, conducted by Paudel Dilli et al. [8] and Wang, Y et al. [9], applied sophisticated techniques in the field of agricultural data analysis and prediction. Paudel Dilli et al. [8] utilized agronomic data combined with ML techniques. The work included information such as crop types, planting dates, soil conditions, irrigation practices, pest and disease management, and other relevant agricultural factors. ML models likely employed regression algorithms, decision trees, neural networks, or other advanced ML techniques. Predictive algorithms were developed to forecast large-scale crop yields based on historical agronomic data. Crop yields were aimed to forecast on a large scale, indicating applicability to broader agricultural contexts. Wang, Y et al. [9] integrated diverse data sources, including crop simulation outputs, weather data, soil data, and remote sensing data. Multiple types of information were combined to create a detailed and holistic view of agricultural conditions. The MCYFS database leveraged the meteorological, crop, and yield forecast system, which contains extensive meteorological data, crop-related information, and yield forecasts. Geographical and crop diversity conducted experiments with various crops across different geographical areas, indicating the versatility of their approach. Both studies demonstrate the importance of integrating various types of data and leveraging AML techniques to enhance agricultural predictions and decision-making processes. Additionally, the use of comprehensive datasets and advanced models highlights the researchers' efforts to create accurate, scalable, and adaptable solutions for the agricultural domain. These approaches contribute significantly to the advancement of precision agriculture and smart farming practices. Wang, Y et al. [9] considered a variety of features including crop simulation outputs, weather data, soil data, and remote sensing data. These features are essential for precision agriculture as they provide crucial information about crop growth conditions, environmental factors, and soil quality. By integrating data from diverse sources, the researchers aimed to create a comprehensive and accurate model for agricultural prediction and decision-making. Additionally, the researchers utilized the MCYFS database for their experiments. The MCYFS database likely contains extensive meteorological data, crop-related information, and yield forecasts. By leveraging this database, the researchers had access to a wealth of historical and real-time data that is fundamental for conducting experiments and validating their models.

A range of machine learning algorithms was employed, including ordinary least squares (OLS), least absolute shrinkage and selection operator (LASSO), support vector machine (SVM), random forest (RF), AdaBoost, and deep neural networks (DNN), to perform comprehensive data analysis. Adaboost method showed the best performance with a reliable prediction ($R2 > 0.84$). While we've mentioned the study conducted by Wang, Y. et al. [9] focusing on the integration of diverse data sources for agricultural predictions, it seems that we're interested in a comparison with the study conducted by Guo, Yahui et al. [10] specifically related to rice yield prediction. We compared traditional multiple linear regression (MLR) with AML methods, including backpropagation (BP), SVM, and RF. We used phenology, climate data, and geographical data as features. ML methods outperformed MLR, with BP, SVM, and RF achieving lower root mean square error (RMSE) and higher R2 values in predicting rice yields.

We've provided an overview of several research papers related to crop prediction and smart farming techniques. Here's a summary of the key points from each of the mentioned papers. Bali et al. [11] conducted a survey of over 80 research papers related to crop prediction. They identified gaps in the existing research and highlighted the use of hybrid models and DL methods like ANN and adaptive neuro-fuzzy inference systems (ANFIS) for crop prediction. These models showed promising results, and the authors suggested exploring further hybridization strategies for improved predictions. Khaki, S. et al. [12] developed a model for forecasting corn and soybean production in the Corn Belt region of the USA from 2016 to 2018. Guo, Yahui et al. [10] employed a diverse set of ml techniques for their rice yield prediction study. These techniques include CNN, which are commonly used for image recognition tasks, but they can also be applied to analyze spatial patterns within agricultural data, such as satellite images or crop field photos. RNN are ideal for sequential data analysis, making them suitable for time-series prediction tasks in agriculture where historical data points are important for making predictions about future yields. RF is an ensemble learning method that combines multiple decision trees to improve accuracy and robustness. It is versatile and can be applied to various types of data, making it suitable for agricultural prediction tasks. Deep feedforward neural networks (DFNN), also known as multilayer perceptron, are a fundamental type of neural network where information moves in only one direction—forward—from the input nodes, through the hidden nodes (if any), and to the output nodes. These networks are effective for capturing complex patterns in data. LASSO is a regression analysis method that performs both variable selection and regularization to enhance the prediction accuracy and interpretability of the statistical model. By employing this combination of techniques, we likely aimed to leverage the unique strengths of each method. CNNs can capture spatial features, RNNs can analyze sequential patterns, RF can handle diverse data types, DFNNs can model complex relationships, and LASSO can perform feature selection and regularization, enhancing the overall prediction accuracy and interpretability of the models. Thus, these resulted in successful predictive models for rice yield, as evidenced by the low RMSE values. Achieving low RMSE values is indicative of the accuracy of the predictions made by the models.

Zhao, J. C. [13] discussed the significance of the IoT in the agricultural industry. They emphasized the integration of advanced technologies, such as control networks and information networks, for remote monitoring, control, and information management in agriculture. Varghese, R., & Sharma, S. [14] focused on leveraging IoT technology and ML for smart farming. They used data collected from various sensors deployed across a farm to predict the future state of crops. This approach aimed to optimize farming practices and resource allocation. Chu, Z., & Yu, J. [15] developed a sorghum yield prediction model using ML techniques. This work uses TensorFlow, CNN, and linear regression to detect sorghum ears in images and estimate their weights. The authors accomplished an average accuracy rate of 74.5% and a precision level of 99% for weight estimation.

Chiu, M. T. et al. [16] aimed to analyze agricultural data using advanced techniques. They utilized a comprehensive dataset containing crucial parameters like precipitation, temperature, evapotranspiration, area, production, and yield. The researchers employed various data mining methods, including K-means clustering, SVM, k-nearest neighbors (KNN), and Bayesian networks, to gain insights and enhance agricultural analysis based on the provided dataset. Dholu, M., & Ghodinde, K. A., in their research on IoT for precision agriculture [17], developed a system focused on precision agriculture applications. They utilized sensor data, including soil moisture, temperature, and relative humidity around plants, to make optimal control decisions. Shinde, S. S., & Kulkarni, M., in their work on crop disease prediction models [18], conducted a comprehensive review of various crop disease prediction models employing ML algorithms. They analyzed environmental parameters such as humidity, temperature, rainfall, wind flow, light intensity, and soil pH to enhance precision agriculture and productivity. The study emphasized the integration of ML models with IoT, highlighting their potential to improve smart farming decisions and control.

Additionally, Park, H. et al. [19] developed an image-based DL model for crop disease prediction where the application is the model to enhance productivity through dynamic image analysis for agriculture decision-making. Kitpo et al. [20] recommended an IoT-DL ecosystem for greenhouse monitoring and control. Parameters are the system-assessed temperature, humidity, water supply, and disease detection for better growth monitoring. Sarangdhar, A. A., & Pawar, V. R. [21] defined leaf disease detection and control with ML regression-based models and IoT. The model is soil quality mentoring model where they applied an SVM-based regression model for classifying cotton leaf diseases. They achieved the accuracy where the maximum classification accuracy achieved was 83.26%. Sharma, S. et al. [22] initiated an objective where the authors proposed a big data analytics model for crop prediction. Input Data as the key information such as average temperature, average humidity, and total rainfall were used for yield prediction. The SVM algorithm was used as a classifier for binary classification of crop yield as good or bad.

These studies collectively demonstrate the wide range of applications of data-driven techniques in agriculture, from predicting crop yields and diseases to optimizing farming practices through IoT and DL. These technologies hold significant potential for improving agricultural productivity and decision-making. As a classifier, the authors used the SVM algorithm to make the two-class classification. The ML has potential to enhance growth in the agriculture sector. Inyaem, U. et al. [23] proposed an ML model, using decision tree and ANN techniques for predicting rice crop production. They also employed N-fold cross-validation to improve accuracy. Shakoor, M. T. et al. [24] applied decision tree learning with ID3 and K-nearest neighbors regression (K-NNR) algorithms to create a prediction model for agriculture production in Bangladesh. This work considers multiple major crop varieties and use in the realm of smart agriculture; several research studies have delved into developing advanced ML models to aid in decision-making processes. Yahata, S., et al. [25] have concentrated their efforts on creating a hybrid model for intelligent agricultural decisions. This approach involves integrating image sensing technologies to monitor soybean flowers and seedpods. Utilizing a combination of sensors, weather data, and various factors, they optimized agricultural choices. The core of their model lies in CNN, which processes the collected data. Materne, N., & Inoue, M. [26] have explored the importance of IoT in a cloud-based setting for predicting agricultural pests and diseases. Leveraging Amazon ML cloud services, they employed a logistic regression (LR) model to unveil hidden patterns and forecast diseases effectively. Chopda, J. et al. [27] proposed a predictive model for cotton crop diseases in the context of smart farming. Their innovative approach incorporates a decision tree classifier alongside IoT and data mining technologies. This integration allows for the prediction of crop diseases, enhancing agricultural practices significantly. Gertphol, S. et al. [28] have contributed to the field by presenting a predictive model tailored for smart hydroponic lettuce farms. Their research involves real-time environmental data collection through IoT sensing technology. Various classification methods, including LR, SVR, multiple MLR, and ANN, were evaluated. Key environmental features impacting plant growth, such as light intensity, humidity, and temperature, were considered. The model's performance was assessed using the root mean square error (RMSE) metric, ensuring accurate evaluation of the predictions made by different models.

Issad, H. A. et al. [29] introduced a smart farming system that harnessed the power of Bayesian networks (BN) and precise inference techniques. This system integrated IoT sensors to continuously gather real-time data on various essential environmental parameters crucial for agricultural practices. These parameters included illuminance, pH levels, electrical conductivity, water temperature, and moisture content. The model utilized BN to make informed inferences about optimal values for these parameters, guided by the data collected from the sensors. In a similar vein Huang, Y. [30] developed a model specifically aimed at estimating soil moisture levels to optimize agricultural practices. This soil moisture estimation model likely employed a combination of diverse techniques and sensors to accurately gauge the moisture levels in the soil. Such information is pivotal for efficient irrigation and overall agricultural management, contributing to improved crop yields and sustainability.

In a specific study conducted by Shahhosseini, M. et al. [31], they employed various ML algorithms, including LASSO regression, ridge regression, RF, and extreme gradient boosting (XGBoost) as meta-models. The main objective of this work is to predict pre-growing season maize yield and nitrate loss. The dataset encompassed experimental data from the Sustainable Corn CAP Research Database, soil information from the SSURGO database, and daily weather data spanning from 1987 to 2016, obtained from Dayment.

Among the algorithms, XGBoost emerged as the most effective, outperforming others with an RMSE of 13.4 percent. This research sheds light on the transformative potential of advanced computational techniques in enhancing agricultural productivity and sustainability. In the second study conducted by Kayad, A. et al. [32], a pioneering approach for monitoring within-field fluctuations in corn yield was proposed. The work employed cutting-edge ML models, including MR, RF, and SVM, to predict these yield variations. The study, spanning the years 2016 to 2018, was carried out on a 22-hectare field.

## 12.3 PROPOSED METHODOLOGY

Table 12.1 likely provides further details about the components, metrics, and performance indicators used in the evaluation and comparison of the models. Overall, it seems to be the core of our research, where we developed and evaluated the HELE model to make predictions in the field of smart farming and intelligent agriculture. It involves a diverse set of ML techniques and emphasizes thorough performance assessment and benchmarking against established methods.

Here's a summary of the key points we've discussed like advanced tools and model evaluation techniques where Weka and Auto-Weka were utilized. In this research work, we harnessed the power of Weka, a versatile Java-based ML package. To enhance the modeling capabilities, the work integrated the Auto-Weka plugin. This integration allowed us to efficiently build a model, modify model parameters, train, assess, and select the most suitable categorization models. Notably, Auto-Weka leverages Bayesian optimization, enabling the automatic optimization of ML model parameters. This sophisticated approach ensures the fine-tuning of our models, enhancing their accuracy and predictive capabilities. Receiver operating characteristic (ROC) curve

*Table 12.1* Components, metrics, and performance indicators

| *Dataset* | *Agriculture datasets from KAGGLE (smart farming, crop-yield production, and plant assessment)* |
|---|---|
| **Pre-processing** | Significant test (Wilkinson test), Min-max normalization outlier analysis |
| **Feature selection** | Rank-sum test, principal component analysis, Pearson correlation analysis, and evolutionary computing-assisted feature selection |
| **Machine learning algorithms** | Decision trees, logistic regression, LOGR, NB |
| **Output** | Predicted output/category |
| **Performance parameters** | Accuracy, precision, F-measure, F1-score, etc. |

(a pivotal tool in assessing the performance of classification models) and model evaluation periodically evaluate our evaluation methodology. The ROC curve visually represents a model's ability to distinguish between classes, with an ideal model's curve approaching y = 1.0. Additionally, the area under the curve (AUC) metric quantifies the ROC curve's performance, with a higher AUC (close to 1.0) indicating superior predictive accuracy. In contrast, random guessing results in an AUC of 0.5. By employing ROC curves and meticulously analyzing AUC values, we gauged the effectiveness of our models, ensuring that our predictions were not only accurate but also robust and reliable.

We collected datasets for five different crops (onion, gram, potato, ragi, and rice) from various districts of Odisha. Data preprocessing involved separating dependent and independent variables and filling missing data with column means. We highlighted the significance of our research, which includes using heterogeneous datasets, which can provide better insights and multiple decision variables for smart farming predictions. We utilized IoT information from benchmark and publicly available sources to make comprehensive analytics more productive for optimal smart farming and intelligent agriculture. Improving data preprocessing, data sampling, feature extraction, and selection enhanced overall computational efficacy and reliability of findings.

### 12.3.1 K-means algorithm

K-means is a widely employed clustering algorithm designed to partition data into K clusters based on their similarities. This algorithm's primary objective is to group data points into clusters in such a way that points within the same cluster are more akin to one another than to those in other clusters. K-means operates on a prototype-dependent partitioned approach, highlighting its nature as a prototype-based clustering algorithm. In practical terms, the algorithm identifies K centroids or prototypes within an n-dimensional space. These centroids act as pivotal reference points, representing the centers of the clusters. By iteratively adjusting these centroids, K-means refines its clusters, ensuring data points are appropriately grouped based on their similarities and dissimilarities. The K-means algorithm begins with the initialization of K centroids (reference points). The number K is determined by the user and represents the desired number of clusters. These initial centroids can be randomly selected from the data points or by using specific initialization strategies. K-means operates in an n-dimensional space, where each data point is represented as a vector of n features or attributes. The algorithm calculates the distance between data points in this space to assign them to the nearest centroid. After initializing the centroids, the K-means algorithm iteratively performs the assignment step, where each data point is allocated to the nearest centroid based on a chosen distance metric (typically Euclidean distance).

The K-Means algorithm clusters data points through an iterative process, refining cluster assignments and centroids to achieve optimal results. Update centroids means initially centroids are positioned randomly. After assigning data points to clusters, the centroids are recalculated as the mean of all data points within each cluster. This step ensures that the centroids accurately represent the cluster centers through iteration. The algorithm iterates through the following steps until convergence. Convergence occurs either when the centroids no longer change significantly or when a predefined number of iterations is reached. Reassignment of data points involves assigning each data point to the nearest centroid, forming clusters based on proximity. Centroids are updated by recalculating the mean of all data points assigned to each cluster. The versatility of the K-means algorithm extends across a wide array of applications, including image segmentation, customer segmentation, and data compression. Its primary strength lies in its ability to reveal natural groupings within datasets. However, the choice of K (the number of clusters) is a crucial parameter. Additionally, different initialization strategies and distance metrics can significantly influence the clustering results, making these considerations pivotal for the algorithm's effectiveness in diverse contexts. K-Means is a widely used clustering algorithm in ML and data analysis for partitioning data into distinct groups. It efficiently divides a dataset into K clusters based on the proximity of data points to cluster centroids. The process begins with the initialization of *K*K, representing the desired number of clusters. The main steps of K-Means begin with the selection of K initial centroids from the dataset. Through iterative clustering the algorithm iterates through the following.

Assigns each data point to the closest centroid, forming K clusters. Convergence occurs when the loop continues until the centroids stabilize, indicating that the clustering process has reached its final state. Proximity functions where K-means utilizes proximity functions like Euclidean distance, Manhattan distance, and Cosine similarity to measure the closeness between data points and centroids Clustering Objective where the algorithm minimizes the Euclidean distance between data points and their assigned cluster centroids, mathematically defining the clustering objective. Outcome prediction for new data points, their distances from existing clusters aid in predicting their outcomes. Cluster sizes in which the resulting cluster sizes, varying with different K values, are analyzed. Notably, clusters with K = 4 are the largest, and those with K = 2 are the smallest. These findings can justify applying feature reduction techniques such as principal component analysis (PCA). In summary, this overview provides insights into K-means, detailing its steps, proximity functions, clustering objectives, outcome predictions, and the significance of cluster sizes in guiding feature reduction methods like PCA.

In the analysis of the training dataset, it is imperative to assess the distribution of 0s and 1s within each cluster. Table 12.2 shows that cluster no. 5 predominantly consists of 0s, comprising 77.92% of the cluster

*Table 12.2* No of 0s and 1s in training set in K-means

| *Cluster no.* | *1* | *2* | *3* | *4* | *5* | *6* | *7* | *8* | *9* |
|---|---|---|---|---|---|---|---|---|---|
| Cluster size | 39 | 28 | 47 | 117 | 96 | 83 | 77 | 105 | 57 |
| Percentage of cluster for 0 | 46 | 73.51 | 67.6 | 71.53 | 77.92 | 70.51 | 68.37 | 54.15 | 76.48 |
| Percentage of cluster for 1 | 43 | 34.489 | 52.42 | 38.46 | 33.07 | 37.48 | 38.62 | 37.84 | 29.51 |

*Table 12.3* No of 0s and 1s in testing set

| *Cluster no.* | *1* | *2* | *3* | *4* | *5* | *6* | *7* | *8* | *9* |
|---|---|---|---|---|---|---|---|---|---|
| Percentage of cluster for 0 | 32.43 | 70 | 70 | 69.24 | 78.5 | 70 | 78.125 | 73.33 | 62.5 |
| Percentage of cluster for 1 | 67.56 | 30 | 30 | 38.46 | 21.5 | 30 | 21.875 | 26.66 | 37.5 |

composition. This abundance of 0s signifies an unproductive yield, indicating that characteristics resembling those within the fifth cluster should be avoided by smallholder farmers. Accuracy plays a pivotal role in our evaluation. Previous projections on the training set revealed that a data point in the first cluster had a 43% chance of yielding a productive harvest. When tested on a different dataset, this probability increased to 67.56%. To gauge the model's accuracy effectively, it is essential to construct a robust loss function. This function will quantify the disparities between the predicted and actual outcomes, thereby providing a measure of the model's predictive power. Additionally, understanding the distribution of 0s and 1s within each cluster is crucial for further refining the model and enhancing its predictive capabilities.

The testing set estimated the quantity of 0s and 1s present in the first nine clusters, as shown in Table 12.3 above.

### 12.3.2 Abnormality

Anomalies in the data became apparent when examining the clusters beyond the first one. The percentages of unproductive yields in both the training and testing sets, as illustrated in Table 12.3, were notably similar. This trend is depicted graphically in Figure 12.1.

Upon establishing the ideal number of K-means clusters, set at 9, it was observed that, excluding the first cluster, the percentages of unproductive

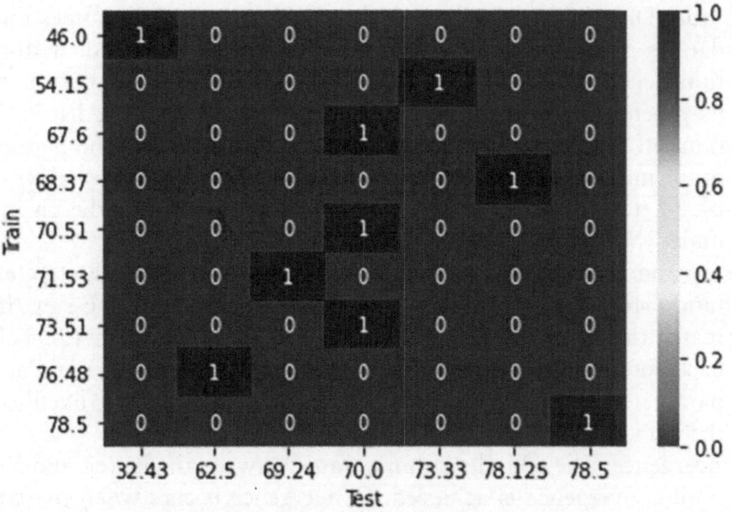


*Figure 12.1* Illustration of K-means clustering results.

yields remained consistent across both the training and testing sets. However, the first cluster exhibited distinctive outcomes in both datasets. This discrepancy underscores the unique characteristics of the first cluster, setting it apart from the rest and highlighting its significance in the analysis. Figure 12.1 illustrates the K-means clustering results. By considering the crop prediction's training and testing in this analysis of crop prediction using training and testing data, this work generated the depicted graph using the K-means clustering algorithms with a specified cluster value of k = 2 and the mean accuracy of 76.48%.

### 12.3.3 Expectation-maximization

The EM algorithm is a statistical technique commonly used for tasks such as parameter estimation and clustering. Let's break down the key points we mentioned as given parameter estimation: EM was originally developed as a method for estimating parameters in statistical models when there are latent (unobserved) variables. It is widely used in cases where you have incomplete data or when the data is generated by a mixture of different probability distributions.

Fuzzy clustering: EM can also be applied to clustering, where it is sometimes referred to as "soft" or "fuzzy" clustering. Unlike K-means, which assigns each data point to a single cluster, EM assigns probabilities to data points belonging to each cluster. This probabilistic approach allows for more flexibility in modeling complex data distributions.

Mixture Distribution: In the context of clustering, EM assumes that the data is generated from a mixture of multiple independent distributions, each representing a cluster. The goal of EM is to estimate the parameters (e.g., mean and variance) of these component distributions.

Maximum Likelihood Estimation (MLE): EM estimates the parameters of the mixture distribution using MLE. It seeks to find the values of the parameters that maximize the likelihood of observing the given data under the assumed mixture model.

E-step and M-step: EM alternates between two steps: the E-step (expectation step) and the M-step (maximization step). In the E-step, it estimates the probability that each data point belongs to each cluster based on the current parameter estimates. In the M-step, it updates the parameters of each cluster to maximize the expected log-likelihood of the data, computed in the E-step.

Convergence: The EM algorithm iterates between the E-step and M-step until convergence is achieved. Convergence occurs when the parameters no longer change significantly or when a predefined number of iterations is reached. Overall, the EM algorithm is a powerful tool for clustering and parameter estimation in various statistical and ML applications. It is particularly useful when dealing with data that may not fit well with simple, rigid cluster assignments, as it provides a more probabilistic and flexible approach to modeling complex data distributions.

### 12.3.4 Linear regression

This work has applied linear regression to a dataset with various features, but we encountered noise in the data. To clean the dataset, we've used various approaches, and then we focused on humidity and precipitation, which are favorably connected to other variables. Following the creation of a graph correlating humidity and precipitation, we computed the mean humidity value. Notably, it was observed that instances with maximum humidity recorded zero precipitation. As a result of these findings, our predictive model achieved an impressive accuracy rating of 89.72 percent. It's great that we were able to improve the accuracy of our model by cleaning the dataset and selecting relevant features. Here are some key points to consider:

Data cleaning: Noise in the dataset can significantly impact the performance of an ML model. It's important to identify and handle missing data, outliers, and any inconsistencies in your dataset. Your mention of "cleaning the dataset using various approaches" suggests that you took steps to address these issues.

Feature Selection: Focusing on relevant features is crucial for improving model accuracy and reducing overfitting. It's good that you chose humidity and precipitation, which you mentioned are favorably connected to other variables. This indicates that you made informed choices regarding feature selection.

Visualization: Visualizing the relationship between features, such as plotting a graph between humidity and precipitation, is a valuable step in understanding your data. It can help you identify patterns and correlations that may not be immediately apparent from summary statistics.

Accuracy: Achieving an accuracy rating of 89.72 percent suggests that your LR model is performing well on the cleaned dataset. However, it's essential to evaluate your model using various metrics, such as mean absolute error (MAE) or RMSE, to get a more comprehensive assessment of its performance.

Zero Precipitation for Maximum Humidity: It's worth noting that having maximum humidity values with zero precipitation might indicate specific weather conditions or data collection anomalies.

Validation: When assessing the performance of your model, it's essential to use validation techniques such as cross-validation to ensure that the reported accuracy is robust and not a result of overfitting to the specific dataset.

$$Yi = \beta 0 + \beta 1xi^1 + \beta 2xi^2 + \beta^3 xi^3 + \beta 4xi^4 \tag{12.1}$$

The following graph illustrates the correlation between humidity and precipitation, highlighting the mean humidity value. Notably, it was observed that instances with maximum humidity recorded zero precipitation. As a result of these findings, our predictive model achieved an impressive accuracy rating of 89.72 percent. Table 12.4 presents the forecast accuracy of the training and testing sets using LR for different crop yields.

Figure 12.2 is plotted by considering humidity and precipitation parameter. The accuracy rate of the potato crop seems to be 95.27 percent, which is high in this case

*Table 12.4* Crop yield forecasts: a comprehensive linear regression approach for diverse crops

| | *Accuracy* | |
|---|---|---|
| *Crops* | *Training set* | *Testing set* |
| Onions | 87.68 | 94.89 |
| Potatoes | 91.39 | 93.79 |
| Gram | 96.49 | 92.61 |
| Ragi | 98.99 | 98.39 |
| Rice | 94.68 | 92.49 |

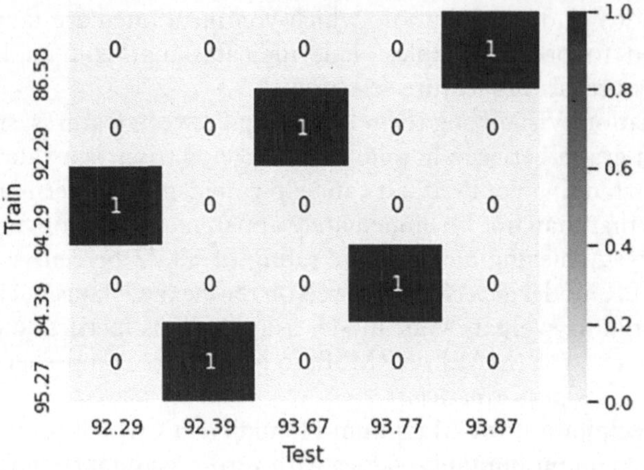


*Figure 12.2* The accuracy of the crop prediction by using logistic regression (LR).

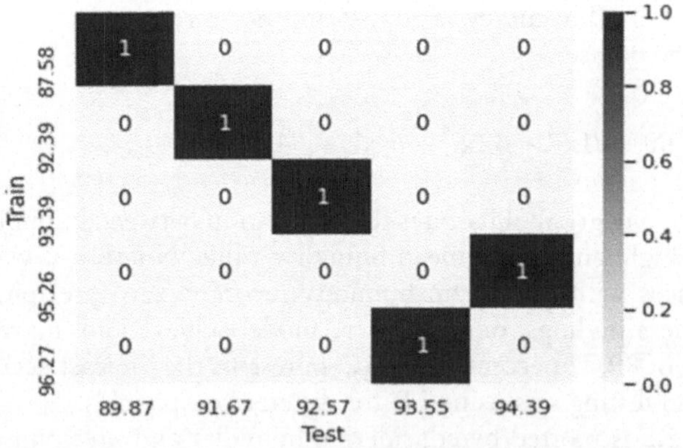


*Figure 12.3* Multifactor predictive modeling using multiple linear regression (MLR).

By considering more than two variables, this work evaluated the accuracy of our hybrid model by using MLR, which unravels complex relationships. Figure 12.3 is plotted between the training and test data used for crop prediction. In this case, the onion crop gives the highest accuracy of 96.27%.

### 12.3.5 Naive bayes

We've provided a detailed explanation of the Bayesian approach to supervised learning and classification. The core concept we are describing is the use of Bayes' theorem in the context of probabilistic classification. Here's a breakdown of the components and equations we've mentioned:

- Bayes' Theorem: Bayes' theorem is a fundamental principle in probability theory and statistics. It allows us to update our beliefs about the probability of an event (in this case, a class label) based on new evidence (input attributes).
- Components of Bayes' Theorem:
  P(A): This represents the prior probability of the response variable (class label).
  P(B): This is the likelihood of the training data or evidence.

In this context, it's denoted as P(B|A), which is the probability of observing the input attributes given a particular class label.

Posterior Probability (P(A|B)): This is the probability of the response variable given the input attributes, which you referred to as the "posterior." It's what you're trying to estimate.

- The Bayes' Theorem Equation: The standard form of Bayes' theorem is given as follows:

$$P(A \mid B) = \left(P(A)^{*} P(B \mid A)\right) / P(B) \tag{12.2}$$

1. Independence Assumption: We mentioned that if A and B are both independent events, then you can use the following equation:

$$P(A,B) = P(A)^{*} P(B) \tag{12.3}$$

2. Predictive Modeling: In the context of classification problems, you're interested in estimating the probability of a class (c) given a set of predictor attributes (x). The posterior probability (P(c|x)) is what you want to predict.
3. Components for Classification:
   P(c): This represents the prior probability of a particular class (c).
   P(x|c): This is the likelihood, which is the probability of observing the predictor attributes (x) given a specific class (c).
   P(x): This is the predictor's prior probability.
4. Bayesian Classification Equation:

$$P(c \mid x) = P(x \mid c)^{*} P(c) / P(x) \tag{12.4}$$

P(c|x): The posterior probability of the class given the attributes, which is what you want to compute.
P(x|c): The likelihood, the probability of the attributes given the class.
P(c): The prior probability of the class.
P(x): The predictor's prior probability.

Absolutely, the equation you provided

$$P(c \mid x) = P(x \mid c)^{*} P(c) / P(x) \tag{12.5}$$

is a direct application of Bayes' theorem to calculate the posterior probability (P(c|x)) of a class (c) given predictor attributes (x). Let's break down the components of this equation again: P(c|x): This is the posterior probability of the class (c) given the observed predictor attributes (x). It's what you want to compute. P(x|c): This represents the likelihood, which is the probability of observing the predictor attributes (x) given a specific class (c). It's a measure of how well the attributes match the characteristics of the class. By applying Bayes' theorem, you can update your belief (prior probability) about the class based on the observed evidence (predictor attributes) to calculate the posterior probability. This is a fundamental concept in probabilistic classification, and it's used in various ML algorithms, including naive Bayes classifiers.

$$P(c \mid x) = P(x|c)P(c) / P(x) \tag{12.6}$$

Table 12.5 presents the accuracy of the different crops. We have estimated the accuracy of the training and testing set and found that the NB algorithm yields a training accuracy of 90.32 percent and just a testing accuracy of 90.12 percent.

In the case of the naive Bayes classifier, the accuracy range varies from 70% to 90%. Figure 12.4 shows that the maximum accuracy for the ragi crop is highest, which is 90.32 %, and for gram is minimum, which is 76.59%.

### 12.3.6 Support vector machine

SVMs are versatile ML algorithms used for both classification and regression tasks. One of their key features is operating in a high-dimensional space,

*Table 12.5* Using naive Bayes to predict different crops

| | *Accuracy* | |
|---|---|---|
| *Crops* | *Training set* | *Testing set* |
| Rice | 84.58 | 86.77 |
| Ragi | **90.32** | **90.12** |
| Gram | 76.59 | 77.77 |
| Potato | 88.22 | 88.19 |
| Onion | 79.77 | 80.25 |

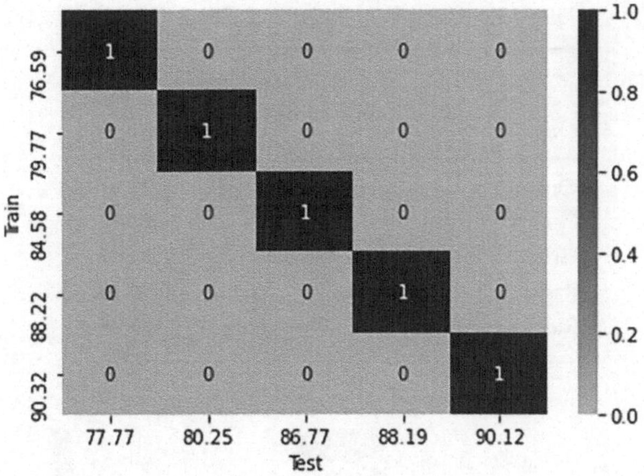


*Figure 12.4* Naive Bayes result analysis.

where each data point is a specific point. In the context of crop prediction, these dimensions represent various crop attributes like soil type, temperature, and rainfall. SVMs excel in categorizing data points. For classification, they seek a hyperplane that maximizes the margin between different classes, ensuring optimal separation. In regression, SVMs find a hyperplane that best fits the data, minimizing errors. The concept of a hyperplane, crucial in SVMs, defines this boundary. To train an SVM, a labeled dataset is utilized, determining the ideal hyperplane for class separation. Testing data, distinct from the training set, assesses how well the SVM generalizes to new, unseen data. In crop prediction, data on attributes such as soil quality and temperature is gathered. A portion trains the SVM, enabling it to learn patterns associated with various crops. Another dataset assesses its ability to correctly predict crop types. In summary, SVMs are potent tools in ML, crafting optimal hyperplanes in high-dimensional spaces. In the realm of crop prediction, they prove invaluable, categorizing crops based on diverse attributes. Rigorous training and testing are essential for evaluating SVM performance in such applications. After training, the SVM's performance is assessed using testing data, which consists of a separate set of data points not used during training. In the realm of crop prediction, SVMs prove their mettle by predicting diverse crops. This involves gathering a wealth of data encompassing critical attributes like soil quality, temperature, humidity, and more. A segment of this data is allocated for training the SVM, enabling it to discern intricate patterns associated with different crop types. Crucially, a separate dataset is reserved for testing, illuminating the SVM's prowess in accurately classifying or predicting crop types. SVMs function by crafting optimal hyperplanes within these multidimensional spaces. These hyperplanes act as intelligent separators, demarcating distinct classes or precisely fitting regression data points. In the context of crop prediction, SVMs emerge

*Table 12.6* Using SVM to predict different crops

| | *Accuracy* | |
|---|---|---|
| *Crops* | *Training set* | *Testing set* |
| Onion | 82.45 | 93.41 |
| Potato | 94.86 | 93.97 |
| Gram | 98.79 | 95.67 |
| Ragi | 94.27 | 90.19 |
| Rice | 98.48 | 72.35 |

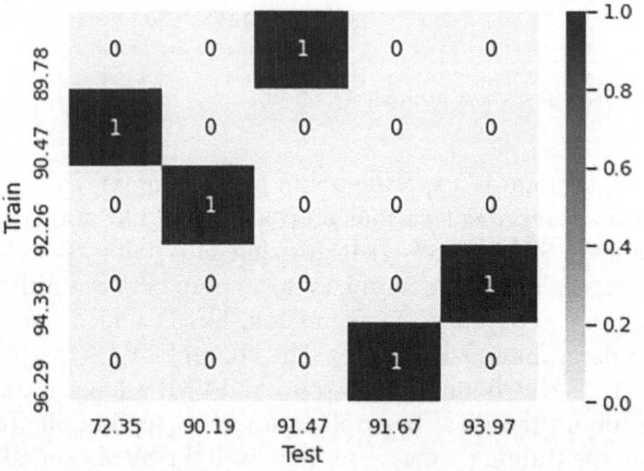


*Figure 12.5* Support vector machine to predict different crops.

as invaluable assets, categorizing crops based on their nuanced characteristics and attributes. Table 12.6 presents the accuracy comparison using SVM.

Figure 12.5 showcases precise predictions for five distinct crop types. After a thorough evaluation of performance metrics, the data strongly advocates for the cultivation of gram crops. This recommendation stems from the impressive accuracy rate of 96.29% achieved through SVM regression analysis. Table 12.8 presents utilization of the SVM to estimate various crops, ultimately leading to the conclusion that gram crops exhibit superior training and testing accuracy (96.29%, 95.67%) compared to other options.

### 12.3.7 Decision tree

The decision tree stands as one of the most commonly employed classification techniques for diverse prediction tasks. Its appeal lies in its ability to create meaningful rules while requiring minimal computational resources. Decision trees offer straightforward guidance by highlighting the essential fields necessary for accurate classification and prediction processes.

*Table 12.7* Employing the decision tree method for crop prediction

| | Accuracy | |
|---|---|---|
| *Crops* | *Training set* | *Testing set* |
| Onion | 97.48 | 99.82 |
| Potato | 96.49 | 99.88 |
| Gram | 99.29 | 99.81 |
| Ragi | 95.28 | 99.89 |
| **Rice** | 99.87 | 99.85 |

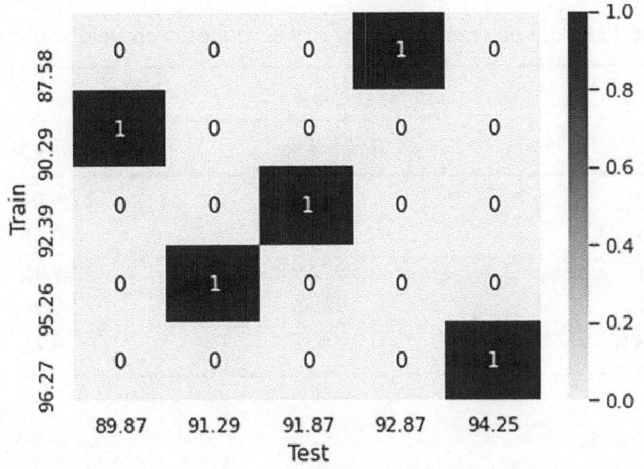


*Figure 12.6* Decision tree.

In Table 12.7, the data was analyzed using decision tree construction, revealing that onions exhibited the highest accuracy rate at 96.27% during the estimation phase. Furthermore, during testing, onions maintained a commendable accuracy rate of 94.25%.

Within the decision tree classifier scenario, the onion crop emerges as the top performer with an impressive accuracy of 96.27%. Figure 12.6 illustrates the prediction of both categorical and continuous data, emphasizing precision in value prediction through classification. To enhance our model's efficiency, a pruning mechanism has been employed.

### 12.3.8 Random forest tree

RFR stands out as a supervised ML technique designed to tackle both classification and regression challenges, aiming to enhance prediction accuracy. Within the RF framework, each tree generates class predictions, ultimately determining the category with the highest prediction score.

Table 12.8 showcases the prediction outcomes obtained through RF classifiers. Based on the algorithm's effectiveness, it is evident that the onion stands out with an exceptional accuracy rate of 99.27%, as demonstrated by the analysis of both training and testing data.

In Table 12.9, the correctly classified instances are 100%, and none of them are incorrectly classified.

Table 12.10 provides a comprehensive breakdown of accuracy by class achieved through RF trees, showcasing various accuracy measurements conducted.

*Table 12.8* Utilizing random forest regression for crop yield prediction

| | *Accuracy* | |
|---|---|---|
| *Crops* | *Training set* | *Testing set* |
| Potato | 98.99 | 99.89 |
| Gram | 99.49 | 99.88 |
| Ragi | 97.69 | 99.86 |
| Rice | 98.46 | 99.99 |
| Onion | 99.27 | 98.25 |

*Table 12.9* Statistical measurement for random forest tree

| *Correctly classifid instances* | *Incorrectly classified instances* | *Kappa statistic* | *Mean absolute error* | *Root mean squared error* | *Relative absolute error* | *Root relative squared error* |
|---|---|---|---|---|---|---|
| 100% | 0% | 1 | 0 | 0 | 0 | 0 |

*Table 12.10* Detailed accuracy by class using random forest tree

| *TP rate* | *FP rate* | *Precision* | *Recall* | *F-measure* | *MCC* | *ROC Area* | *PRC area* |
|---|---|---|---|---|---|---|---|
| 1.00 | 1.00 | 1.00 | 1.00 | 1.00 | 1.00 | 1.00 | 1.00 |
| 1.00 | 1.00 | 1.00 | 1.00 | 1.00 | 1.00 | 1.00 | 1.00 |
| Weighted Avg 1.00 | 1.00 | 1.00 | 1.00 | 1.00 | 1.00 | 1.00 | 1.00 |

### 12.3.9 Neural network

We've provided some information about ANNs and RNNs. We'll provide some additional context and clarify a few points:

1. Artificial Neural Network:
   ANNs are indeed a powerful learning strategy inspired by the human brain's neural structure.
   They consist of interconnected artificial neurons that process information.
   ANNs can learn from previous experiences or examples, making them useful for various tasks, including pattern recognition, classification, and regression.
   ANNs require training data to perform effectively. The quality and representativeness of the training data are critical to their success.
   One common issue with ANNs is the lack of interpretability. It can be challenging to understand how the network arrives at its decisions, especially in complex models.
2. Recurrent Neural Network:
   RNNs are a specific type of ANN designed for handling sequential data, making them suitable for tasks involving time series, natural language processing (NLP), and speech recognition. RNNs have a memory element that allows them to maintain information over short-term sequences. This makes them suitable for tasks involving dependencies within sequential data. However, traditional RNNs have limitations in retaining information over long sequences. This is due to the vanishing gradient problem, which can cause information loss when dealing with lengthy sequences. To address the vanishing gradient problem, more advanced RNN variants like LSTM and gated recurrent unit (GRU) have been developed. These variants can capture longer-term dependencies in data.

In summary, ANNs and RNNs are valuable tools in ML and AI. ANNs can handle a wide range of tasks but require careful training and may lack interpretability. RNNs, including LSTM and GRU variants, excel at handling sequential data and have been instrumental in advancing NLP and other time-series tasks, mitigating some of the issues associated with traditional RNNs.

$$\text{The equation } \big(h(t)\big) = fc\big(h(t-1)\big) \tag{12.7}$$

(x(t)) represents the calculation of a new state, denoted as (h(t)). In this equation, (fc) represents a function with parameter (c), incorporating the previous state (h(t – 1)) and the input vector (x(t)) at time step (t). Table 12.11 presents the test accuracy across different hidden layers using various activation functions.

*Table 12.11* Contrasts different activation functions and their test accuracy across hidden layers

| *Hidden layer* | *Activate function* | *Hidden neurons* | *Test accuracy* |
|---|---|---|---|
| 2 | Sigmoid | 6 | 90.53% |
| 3 | Sigmoid | 7 | 92.85% |
| 4 | Sigmoid | 8 | 99.71% |
| 5 | Sigmoid | 11 | 93.71% |
| 2 | ReLU | 6 | 80.47% |
| 3 | ReLU | 7 | 82.46% |
| 4 | ReLU | 8 | 85.68% |
| 5 | ReLU | 11 | 82.72% |
| 2 | Tanh | 6 | 69.34% |
| 3 | Tanh | 7 | 71.86% |
| 4 | Tanh | 8 | 77.5% |
| 5 | Tanh | 11 | 78.29% |

## 12.4 CONCLUSION

In conclusion, this study has tackled a crucial issue faced by Indian farmers—the selection of the most suitable crop for their agricultural needs. Leveraging ML techniques, a recommendation system was developed to provide crop suggestions based on geographical and climatic factors. The system encompasses data on five distinct crops: rice, ragi, gram, potato, and Onion. Various ML and deep learning methods were employed, with the RF model achieving an impressive accuracy of 99.27% on the training set and 98.25% on the testing set, indicating a minimal 1.02% difference between the two. Furthermore, in a comparison with activation functions like sigmoid, ReLu, and tanh, sigmoid outperformed the others, reaching a peak accuracy of 99.71% with four hidden layers. This study reinforces the efficacy of RF models for crop prediction, offering a promising solution to assist Indian farmers in making informed crop selection decisions.

## REFERENCES

[1] Van Klompenburg, T., Kassahun, A., & Catal, C. (2020). Crop yield prediction using machine learning: A systematic literature review. *Computers and Electronics in Agriculture*, *177*, 105709.

[2] Wilson, G. L., Mulla, D. J., Vetsch, J. A., & Sands, G. R. (2020). Predicting nitrate-nitrogen loads in subsurface drainage as a function of fertilizer application rate and timing in southern Minnesota. *Journal of Environmental Quality*, *49*(5), 1347–1358.

[3] Kok, Z. H., Shariff, A. R. M., Alfatni, M. S. M., & Khairunniza-Bejo, S. (2021). Support vector machine in precision agriculture: a review. *Computers and Electronics in Agriculture*, *191*, 106546.
[4] Tian, Y., Xu, Y. P., & Wang, G. (2018). Agricultural drought prediction using climate indices based on Support Vector Regression in Xiangjiang River basin. *Science of the Total Environment*, *622*, 710–720.
[5] Egmont-Petersen, M., de Ridder, D., & Handels, H. (2002). Image processing with neural networks—a review. *Pattern Recognition*, 35(10), 2279–2301.
[6] Jin, X., Che, J., & Chen, Y. (2021). Weed identification using deep learning and image processing in vegetable plantation. *IEEE Access*, *9*, 10940–10950.
[7] Apat, S. K., Mishra, J., Srujan Raju, K., & Padhy, N. (2022). State of the art of ensemble learning approach for crop prediction. *Next Generation of Internet of Things: Proceedings of ICNGIoT 2022*, 675–685.
[8] Paudel, D., Boogaard, H., de Wit, A., Janssen, S., Osinga, S., Pylianidis, C., & Athanasiadis, I. N. (2021). Machine learning for large-scale crop yield forecasting. *Agricultural Systems*, *187*, 103016.
[9] Wang, Y., Zhang, Z., Feng, L., Du, Q., & Runge, T. (2020). Combining multisource data and machine learning approaches to predict winter wheat yield in the conterminous United States. *Remote Sensing*, *12*(8), 1232.
[10] Guo, Yahui, et al. Integrated phenology and climate in rice yields prediction using machine learning methods. *Ecological Indicators* 120 (2021): 106935.
[11] Bali, N., & Singla, A. (2022). Emerging trends in machine learning to predict crop yield and study its influential factors: A survey. *Archives of Computational Methods in Engineering*, *29*(1), 95–112.
[12] Khaki, S., Wang, L., & Archontoulis, S. V. (2020). A CNN-RNN framework for crop yield prediction. *Frontiers in Plant Science*, *10*, 1750.
[13] Zhao, J. C., Zhang, J. F., Feng, Y., & Guo, J. X. (2010, July). The study and application of the IOT technology in agriculture. In *2010 3rd International Conference on Computer Science and Information Technology* (Vol. 2, pp. 462–465). IEEE.
[14] Varghese, R., & Sharma, S. (2018, June). Affordable smart farming using IoT and machine learning. In *2018 Second International Conference on Intelligent Computing and Control Systems (ICICCS)* (pp. 645–650). IEEE.
[15] Chu, Z., & Yu, J. (2020). An end-to-end model for rice yield prediction using deep learning fusion. *Computers and Electronics in Agriculture*, *174*, 105471.
[16] Chiu, M. T., Xu, X., Wei, Y., Huang, Z., Schwing, A. G., Brunner, R., ... & Shi, H. (2020). Agriculture-vision: A large aerial image database for agricultural pattern analysis. In *Proceedings of the IEEE/CVF Conference on Computer Vision and Pattern Recognition* (pp. 2828–2838).
[17] Dholu, M., & Ghodinde, K. A. (2018, May). Internet of things (IoT) for precision agriculture application. In *2018 2nd International Conference on Trends in Electronics and Informatics (ICOEI)* (pp. 339–342). IEEE.
[18] Shinde, S. S., & Kulkarni, M. (2017, December). Review paper on prediction of crop disease using IoT and machine learning. In *2017 International Conference on Transforming Engineering Education (ICTEE)* (pp. 1–4). IEEE.
[19] Park, H., JeeSook, E., & Kim, S. H. (2018, August). Crops disease diagnosing using image-based deep learning mechanism. In *2018 International Conference on Computing and Network Communications (CoCoNet)* (pp. 23–26). IEEE.

[20] Kitpo, N., Kugai, Y., Inoue, M., Yokemura, T., & Satomura, S. (2019, January). Internet of things for greenhouse monitoring system using deep learning and bot notification services. In *2019 IEEE International Conference on Consumer Electronics (ICCE)* (pp. 1–4). IEEE.

[21] Sarangdhar, A. A., & Pawar, V. R. (2017, April). Machine learning regression technique for cotton leaf disease detection and controlling using IoT. In *2017 International Conference of Electronics, Communication and Aerospace Technology (ICECA)* (Vol. 2, pp. 449–454). IEEE.

[22] Sharma, S., Rathee, G., & Saini, H. (2018, December). Big data analytics for crop prediction mode using optimization technique. In *2018 Fifth International Conference on Parallel, Distributed and Grid Computing (PDGC)* (pp. 760–764). IEEE.

[23] Inyaem, U. (2018, June). Construction model using machine learning techniques for the prediction of rice produce for farmers. In *2018 IEEE 3rd International Conference on Image, Vision and Computing (ICIVC)* (pp. 870–874). IEEE.

[24] Shakoor, M. T., Rahman, K., Rayta, S. N., & Chakrabarty, A. (2017, July). Agricultural production output prediction using supervised machine learning techniques. In *2017 1st International Conference on Next Generation Computing Applications (NextComp)* (pp. 182–187). IEEE.

[25] Yahata, S., Onishi, T., Yamaguchi, K., Ozawa, S., Kitazono, J., Ohkawa, T., Yoshida, T., Murakami, N. & Tsuji, H. (2017, May). A hybrid machine learning approach to automatic plant phenotyping for smart agriculture. In *2017 International Joint Conference on Neural Networks (IJCNN)* (pp. 1787–1793). IEEE.

[26] Materne, N., & Inoue, M. (2018, March). IoT monitoring system for early detection of agricultural pests and diseases. In *2018 12th South East Asian Technical University Consortium (SEATUC)* (Vol. 1, pp. 1–5). IEEE.

[27] Chopda, J., Raveshiya, H., Nakum, S., & Nakrani, V. (2018, January). Cotton crop disease detection using decision tree classifier. In *2018 International Conference on Smart City and Emerging Technology (ICSCET)* (pp. 1–5). IEEE.

[28] Gertphol, S., Chulaka, P., & Changmai, T. (2018, November). Predictive models for lettuce quality from internet of things-based hydroponic farm. In *2018 22nd International Computer Science and Engineering Conference (ICSEC)* (pp. 1–5). IEEE.

[29] Issad, H. A., Aoudjit, R., & Rodrigues, J. J. (2019). A comprehensive review of Data Mining techniques in smart agriculture. *Engineering in Agriculture, Environment and Food*, *12*(4), 511–525.

[30] Huang, Y. (2023). Improved SVM-based soil-moisture-content prediction model for tea plantation. *Plants*, *12*(12), 2309.

[31] Shahhosseini, M., Martinez-Feria, R. A., Hu, G., & Archontoulis, S. V. (2019). Maize yield and nitrate loss prediction with machine learning algorithms. *Environmental Research Letters*, *14*(12), 124026.

[32] Kayad, A., Sozzi, M., Gatto, S., Marinello, F., & Pirotti, F. (2019). Monitoring within-field variability of corn yield using Sentinel-2 and machine learning techniques. *Remote Sensing*, *11*(23), 2873.

Chapter 13

# Drones

## The unmanned aerial vehicles for precision management of pests and diseases of crops

*Namburi Karunakar Reddy, Pooja Purushotham, Shivani Jha, Gaurav Y. Rakhonde, Naide Prathap Reddy and Somireddy Srinivasa Reddy*

### 13.1 INTRODUCTION

Recently, realm of agricultural and environmental administration has witnessed a transformative technological leap through integration of drones, also referred to as UAVs. These versatile and remotely operated devices have emerged as game-changers in the field of precision pest and disease management. Through their capacity to soar above fields and landscapes with unprecedented ease and agility, drones have revolutionized the approach and challenges of monitoring, identifying, and mitigating pests and diseases that can wreak havoc on crops and ecosystems.

This paradigm shifts from traditional methods to drone-based precision management represents a pivotal advancement that holds the promise of enhanced efficiency, reduced environmental impact, and optimized resource utilization. By leveraging cutting-edge sensors, imaging technologies, and data analytics, drones empower agriculturalists, environmentalists, and researchers to meticulously survey vast expanses of land, pinpointing areas afflicted by pests and diseases with unparalleled accuracy. Moreover, drones facilitate rapid response interventions, enabling targeted treatments that minimize the use of chemicals and resources, thus promoting sustainability and safeguarding biodiversity.

This exploration of UAVs in precise pest and disease management delves into their diverse applications, the technical advancements enhancing their effectiveness, the benefits they provide, and the challenges that need to be addressed. By harnessing the synergy of aviation and data-driven insights, we will embark on a journey that redefines our approach to pest and disease management, ushering in a new era of innovation and sustainable stewardship in agriculture and beyond.

DOI: 10.1201/9781003628392-13

## 13.2 TYPES OF DRONES

In the dominion of precision management of pests and diseases, various types of drones are employed to address specific challenges and tasks. These drones possess specialized sensors, cameras, and technologies enabling us to gather data, monitor conditions, and deliver targeted interventions. Here are some common types of drones used in the precision management of pests and diseases:

**Multi-rotor drones:** These are versatile drones equipped with multiple rotors, typically quadcopters or hexacopters. They offer stability, maneuverability, and the ability to hover in place, making them ideal for close-up inspections, mapping small- to medium-sized fields, and capturing high-resolution images.

**Fixed-wing drones:** These resemble small airplanes and are designed for longer flight durations and broader coverage. They excel at covering large areas efficiently, making them suitable for mapping expansive agricultural landscapes and tracking the spread of pests and diseases over vast regions.

**VTOL (Vertical Takeoff and Landing) drones:** VTOL drones possess features of both fixed-wing and multi-rotor drones. They can take off as well as land vertically like a quadcopter but transition to horizontal flight like an airplane. VTOL drones are well suited for scenarios where runway space is limited, and they offer flexibility in both maneuverability and coverage.

**Hybrid drones:** These drones merge capabilities of different drone types to provide enhanced versatility. For instance, a hybrid drone might combine the vertical takeoff and landing capabilities of a multi-rotor drones with the endurance of a fixed-wing drone for longer flights and wider coverage.

**Spraying drones:** These drones are equipped with liquid or granular spraying systems. They can precisely deliver pesticides, herbicides, and other treatments to targeted areas, reducing the need for widespread chemical application and minimizing environmental impact.

**Remote sensing (RS) drones:** These are with specialized sensors such as multispectral, hyperspectral, and thermal cameras, LiDAR (Light Detection and Ranging), and more; these drones can capture various data types beyond visible light. RS drones enable the detection of stress, disease symptoms, and other subtle changes in crops that might not be visible to the naked eye.

**Autonomous drones:** Many modern drones are equipped with advanced autonomous capabilities. They can follow pre-programmed flight paths, automatically avoid obstacles, and carry out missions without constant manual control. This autonomy enhances efficiency and reduces the cognitive load on operators.

**Data analysis drones:** These drones are focused on collecting data for analysis rather than physical interventions. These drones have high-resolution cameras and sensors. It captures data to perceive initial signs of disease outbreaks, pest infestations, and stress in crops.

**Bio-control release drones:** In cases where biological agents (predators, parasitoids, etc.) are used to control pests, drones can be employed to release these agents over targeted areas with precision, improving the effectiveness of biological pest management.

The selection of drone is based on specific purpose of task at hand, scale of operation, terrain, and technologies required to gather accurate data or deliver precise interventions. The integration of these diverse drone types represents a remarkable advancement in pest and disease management, offering tailored solutions to enhance agricultural productivity and environmental sustainability.

## 13.3 ADVANTAGES OF DRONES OVER CONVENTIONAL METHODS

The utilization of drones offers a range of advantages over conventional methods. These advantages stem from the unique capabilities and technologies that drones bring to the table. Here are some key benefits:

**High-resolution imaging:** Drones have the advanced cameras equipped with sensors to capture high-resolution images and data from various spectral bands. This allows for early detection of subtle changes in plant health, enabling timely interventions before pest or disease outbreaks become severe.

**Rapid and comprehensive surveillance:** Drones are efficient to cover large areas quickly and efficiently. This rapid surveillance potential is valuable for timely detection of pests and diseases, allowing farmers to respond promptly and prevent the spread of infestations.

**Flexibility and accessibility:** Drones have good ability to access remote or hard-to-reach areas that might be exciting for ground-based methods. They can fly over rugged terrain, water bodies, and dense vegetation, providing a comprehensive view of the entire landscape.

**Reduced labour and time:** Traditional scouting methods often involve manual labour and time-consuming processes. Drones automate these tasks, saving labour costs and freeing up valuable time for farmers and researchers to focus on other essential activities.

**Precision application of treatments:** Drones equipped with spraying systems can deliver pesticides and other treatments with high precision. This targeted application reduces the amount of chemicals used, minimizes environmental impact, and prevents overexposure to non-target areas.

**Data-driven insights:** Drones collect a wealth of data that can be analysed to generate valuable insights. From crop health assessments to pest distribution patterns, this data enables informed decision-making and helps tailor interventions for maximum effectiveness.

**Reduced environmental impact:** Drones enable more precise and localized application of treatments. This leads to reduce overall consumption of chemicals, which in turn mitigates negative impacts on the environment and non-target species.

**Real-time monitoring:** The drones deliver real-time data, agreeing for quick adjustments in management strategies based on changing conditions. This adaptability is crucial for addressing dynamic pest and disease scenarios.

**Cost-effectiveness:** While the initial investment in drone technology might be substantial, long-term benefits, including labour savings, reduced chemical use, and increased yield, can lead to cost savings over time.

**Scalability:** Drones can be deployed on numerous scales, from small individual farms to huge commercial operations. This scalability ensures that the technology is accessible and adaptable to different agricultural contexts.

**Safety:** By conducting aerial surveys and interventions, drones can reach the potentially hazardous environments, such as areas with pesticide exposure or uneven terrain where human expose can be fatal.

**Data integration and analysis:** Drones can be integrated to increase efficiency with Geographic Information Systems (GIS) and other data management platforms, allowing for the creation of detailed maps, trend analysis, and predictive modelling optimizing pest and disease management strategies.

Drones offer a transformative approach to precision management of pests and diseases, enhancing efficiency, accuracy, and sustainability in agricultural practices. By leveraging their capabilities, farmers, researchers, and environmentalists can make informed decisions, minimize losses, and contribute to more resilient and productive food systems.

## 13.4 DRONES IN PRECISION AGRICULTURE PEST AND DISEASE MANAGEMENT

The United Nations Food and Agriculture Organization (FAO) has recently indicated in a report that over 815 million individuals are facing chronic hunger, with Asia representing 64 per cent. Meeting the food demands of a projected nine billion people by 2050 requires an estimated 50 per cent increase in global food production (Ehrlich and Harte, 2015). Many farmers were unable to obtain elementary agricultural necessities such as labour, seeds, fertilizers, and pesticide application on time, resulting in lower production

(Varshney *et al.*, 2020; Stephens *et al.*, 2020). Farmers use traditional methods for seed planting, composting, and pesticide application, among other things. Traditional ways are time-consuming and less efficient, prompting necessity advancements in this field (Mrema *et al.*, 2014). COVID-19 pandemic exacerbated challenges for conventional farmers in monitoring crop care and fertilizer and pesticide application (Varshney *et al.*, 2020; Hobbs, 2020). The adoption of drones in agriculture serves as a viable solution to overcome these challenges (Radoglou-Grammatikis *et al.*, 2020), offering valuable data to agronomists, rural specialists, and farmers for optimizing their operations. Any further decreases in food source in future will result in a terrible condition in the world as crises are arising due to challenging climate change. UAVs are furnished with cameras, sensors, and sprayers to monitor crops and apply pesticides. Yamaha created initial UAV models for agriculture (Giles and Billing, 2015). A scientific ideology of UAV application in precision agriculture guises at how well they can be employed for tasks like crop monitoring, crop height valuation, spraying pesticides, and soil and field analysis (Primicerio *et al.*, 2012; Bendig *et al.*, 2012; Huang *et al.*, 2009; Anthony *et al.*, 2014). The combination of a UAV with sprayer creates an aerial platform for vector and pest management.

### 13.4.1 Components of drone-mediated sprayer system

Electronic speed control (ESC), brushless direct current motors (BLDC), cameras, transmitters, flight controllers, and receivers are key elements of drone. The pump and its control system are two essential components of any spraying system. The drone is controlled using an accelerometer, gyroscope, and GPS. The first step in designing a drone for spraying applications is to estimate the payload. After calculating the payload, elements of drone are chosen. The battery type is determined by current and voltage essential to drone modules. Drone's frame is premeditated based on the number of arms and payloads. Crop monitoring and pesticide spraying are two precision farming activities that require an aerial image processing system. The system's image processing efficiency is determined by image capture systems, aerial platforms, and machine learning (ML) algorithms. The UAV sprinkling system through GPS coordinates automatically spray pesticides on infested spots where no vegetation was perceived by Normalized Difference Vegetation Index (NDVI). This reduces wastage of water and chemicals and also its drifts.

This location-specific application is appropriate for huge agricultural land. Heavy-lift UAVs (Sarghini and De Vivo, 2017a; Sarghini and De Vivo, 2017b) are necessary for this use in order to spray an extensive region. Through the use of PWM (pulse width modulation) controller, the spraying system attached to UAV works much efficiently when applying pesticides (Huang *et al.*, 2009; Zhu *et al.*, 2010). Deposition of pesticides from established UAV is efficient similar to ground-based sprayers. Due to their accuracy and speed, UAVs are gaining attention in spraying operations. However, some factors, such as improper coverage of some crop fields during spraying, crop areas that overlap, and spraying at the crop field's edges, decrease the superiority of harvest.

To address these challenges, a group of UAVs (Yao *et al.*, 2016) were implemented in a control circlet of an algorithm for agricultural operations, where UAVs spray pesticides. The response from Wireless Sensor Network (WSNs) executed in field organizes the process of applying pesticides on crop (Costa *et al.*, 2012). Each control loop communicates with UAVs, adjusting its route based on deviations in wind speed and the number of exchanged messages (Faical *et al.*, 2014). A brief delay in control loop enables UAVs to analyse data from the WSN and identify the optimal route. It may also reduce pesticide waste (Kale *et al.*, 2015).

### 13.4.2 Advantages of drone-mediated spraying pesticides

Advancements in UAV technology offer practical solutions in scenarios where human intervention is impractical, particularly in addressing labour shortages during spraying on crops. These technological enhancements not only streamline spraying process but also enhance efficiency, involving crop monitoring *via* multispectral camera mounted on a UAV. During a single flight, the camera captures images and employs geographic indicators for sophisticated analysis, facilitating the precise determination of pesticide application areas. The continual reduction in size and cost of drones has rendered them ubiquitous in crop health monitoring and livestock detection. Unmanned helicopters, owing to their substantial payload capacity, are primarily employed for pesticide or fertilizer spraying. Additionally, multicopters are gaining prominence in pesticide application due to their superior flight stability, making them ideal for targeted spraying. Noteworthy progress in drone camera technology, including reductions in weight, size, and enhanced resolution, is evident. The transition from red-green-blue (RGB) to multispectral cameras is driven by heightened feature extraction requirements. Moreover, the integration of ML in drones has yielded a user-friendly system tailored to the needs of farmers.

The utilization of UAVs in precision agriculture is currently in its budding phase, offering occasions for advancements in both technological capabilities and agricultural applications. Anticipated progress in UAV technology holds the promise of enhanced image processing techniques, reduced costs, extended flying times, improved battery technologies, innovative camera designs, low-volume sprayers, vision degradation mitigation, increased end user technology literacy, and diversified nozzle types. Numerous experimental studies focusing on UAV-based remote sensing (RS) for agricultural applications have been shown, indicating the potential of these systems to offer substantial benefits in precision agriculture and environmental monitoring.

## 13.5 DRONE-MEDIATED REMOTE SENSING

RS is both an art and science to learn about something or something happening without actually coming into contact with it. Remote monitoring is presently employed on a broader scale than required for precision agriculture, encompassing applications in soil mapping, terrain analysis, crop stress assessment, yield mapping, and soil organic matter estimations. Due to its ability to track spatial variability, high-resolution RS can be very helpful in precision disease and pest management (Moran *et al.*, 1997). A portion of signal that RS satellites send to the earth and receive is replicated back. Image data is collected actively by measuring either the electromagnetic energy emitted from an object or the solar energy reflected by an object. The resolution, spatial coverage, and frequency of remote sensing can vary; for example, higher resolutions are needed to measure the severity of a disease infestation than are needed to monitor a weed infestation (Chinmay *et al.*, 2007).

Drone-based RS technologies have a number of positive aspects appealing for precise pest management. The sensing drones will cover larger areas than handheld with ground-based devices. Particular biotic and abiotic stresses persuade physiological plant responses, alter the plant's ability to perform photosynthesis, and thus alter the spectral range of leaf reflectance.

In aerial RS, a drone will be armed with RGB sensor, a multispectral sensor with 3 to 12 broad spectral bands, or a hyperspectral sensor with hundreds of narrow spectral bands (Keller and Shields, 2014). RS identifies patterns of canopy reflectance, revealing plant stress caused by insect pests rather than the pests themselves. Consequently, on-site field observations still remain essential to confirm THE presence of specific insect pests.

Drone imagery is extremely advantageous in predicting yield loss. RGB values are defined in pixels by RGB cameras, which are frequently used in cameras. These cameras produce images to measure the degree of the intensity of three colours. Crop biomass is estimated using RGB cameras, which helps create 3-dimensional (3D) models in agriculture (Surovy *et al.*, 2018; Torres-Sanchez *et al.*, 2018; Chang *et al.*, 2017). To increase accuracy when calculating the biomass, RGB cameras are equipped with near-infrared (NIR) and multispectral cameras (Roth and Streit, 2018). A modified RGB camera is one that has a red filter in place of NIR filter (Nijland *et al.*, 2014; Berra *et al.*, 2017). RGB cameras are commonly less expensive and are widely accessible. They can also be used to gather crisp still pictures. However, they can only measure the electromagnetic spectrum's three red, green, and blue bands. As a result, in terms of the camera system's spectral resolution, RGB images are less precise than multispectral or hyperspectral images. It is impossible to distinguish between different sheath blight severity levels in rice using images from RGB cameras (Zhang *et al.*, 2018). However, compared to multispectral systems, the main compensations of RGB cameras are their capability to capture high spatial resolution images, which provides more spatial particulars for plant disease discovery and monitoring. RGB cameras must be employed judiciously to ensure that the images have consistent lighting and colouring. Less error will be visible in uniform images when identifying healthy and diseased plants.

## 13.6 DRONE-MEDIATED AERIAL PHOTOGRAPHY AND HYPERSPECTRAL IMAGING

The convergence of wireless sensors, network clustering, and precision agricultural design has facilitated plant pest surveillance through drone-mediated aerial photography. Drones offer a comprehensive aerial perspective of fields, enabling timely decision-making for effective problem resolution. Captured images are transmitted to a cloud data centre, where spectrum analysis technology, as outlined by Gao *et al.* (2020), is employed to analyse pest damage. Multispectral, hyperspectral aerial and satellite imagery contribute to the generation of NDVI maps. These maps help distinguish between soil and vegetation types, identifying stressed plants, and discerning crop types and stages. The strong correlation between crop yield and NDVI data, particularly measured at specific crop stages (Huang *et al.*, 2013), underscores the significance of tracking crop growth during key phases. This provides insights into the estimation of crop yield and allows for initial identification of issues.

Within agricultural mapping tools, the integration of NDVI data with other indices like CWSI (Crop-Water Stress Index) and CCCI (Canopy-Chlorophyll Content Index) yields valued understandings into crop health. The fundamental concept underlying NDVI is rooted in the fact that leaves exhibit substantial reflection of light in NIR spectrum. When a plant undergoes dehydration or stress, leaves reflect reduced NIR light while maintaining a consistent level of reflection in the visible range. So, mathematically merging these two indicators will distinguish between plants and non-plants, as well as between healthy and diseased plants.

UASs (unmanned aircraft system) are becoming more and more popular in agriculture and forestry because they are a convenient and affordable tool for surveying land and crops, gathering data for further analysis, and assisting in managing and decision-making developments. The commonly utilized sensors, like RGB and multispectral sensors, often offer limited data on a restricted number of bands or spectral ranges. This may prove insufficient for comprehensive data representation of scanned areas, such as forest sections or crop areas. To address, hyperspectral sensors, traditionally employed on satellites or manned aircraft, are undergoing redesign to become lighter and smaller, enabling their integration with UAVs for earth surface surveying.

This redesign incorporates a broader spectral range and narrower bands to enhance the sensors' capability in capturing more detailed and comprehensive data.

Hyperspectral sensors create vast data due to their ability to capture spectra sets comprising hundreds of bands across images with high spatial resolution. Such dimensionality necessitates appropriate processing and analysis techniques. Such tools might be useful for those working on agro-forestry and related applications in fields of RS, land surveying, and programming. Studies using UAVs and hyperspectral sensors in the fields of agriculture and forestry were examined, with a focus on their primary objectives and general approaches to processing and analysis. Even though there are many applications, they seem to be more limited than applications using UAVs along with other types of sensors. This is likely due to high cost of high-resolution spectroscopy to compromise cost efficiency of applications involving agro forestry and other associated fields. The difficulty in data collection and analysis is another factor contributing to a low number of studies. Acquiring the data necessitates a more advanced level of training. Despite predictions by some scientists, such as Salami *et al.* (2014), suggesting that hyperspectral sensors will be limited to research applications, it was challenging to endorse this perspective. The increasing availability of market solutions advocating the extensive adoption of high-resolution spectroscopy devices contradicts this viewpoint. Furthermore, it is anticipated that technological advancement in the coming years will result in smaller, more cost-effective devices, making hyperspectral-based sensing a common practise in agriculture, forestry, and related fields.

Certainly, a gap of chance exists for proliferation of business models centred around decentralized and cost operative service supply, specifically targeting this type of investigation. This approach facilitates the development of more precise decision-support tools. However, it is acknowledged that UAS-based RS using hyperspectral technology is still a considerable distance away from becoming economically accessible for common farmers and foresters. In light of this and in accordance with the proposal already made in (Adao *et al.*, 2017), further work will be done to identify and classify phytosanitary problems to affect vineyards by hyperspectral-based RS carried out by UAVs.

## 13.7 DRONE-MEDIATED INSECT PEST SAMPLING AND DISEASE DETECTION

Various insect sampling methods, both active and passive, are available, ranging from highly selective to capable of collecting a diverse array of species. Common techniques include Malaise, colour pan/bowl (Moericke), interception traps, and light, with traditional hand-sampling using a net bag being particularly versatile (Samways *et al.*, 2010). However, hand-sampling is time-consuming and leaves footprints. In challenging environments like steep slopes, wetlands, or areas with dangerous animals, traditional methods may be impractical. Active sampling becomes crucial in such inaccessible study sites, allowing researchers to overcome limitations and collect taxa that are otherwise difficult to sample. These conditions pose obstacles to taxonomical research, particularly significant in tropical areas where insect diversity is extensive, and numerous species remain undiscovered.

Drones will aid in biodiversity research by giving aerial pictures from remote and are typically insufficient for most entomological investigations; rather, specimen collecting is necessary. An inexpensive standard drone was equipped with a net bag, and flown over ground, sweeping vegetation and sampling adult and larvae insects. Drone-netting emerges as a precise tool for insect sampling, especially in challenging and remote areas. It proves to be both time and cost operative, offering minimal intrusion and flexibility for various exploration activities. Its representativeness is comparable to traditional hand-netting, and the captured specimens, remaining alive, can be released if not required. The exploration of using drones equipped with a net bag, termed drone netting, stems from the success of drones in detecting, monitoring, and managing insect pest populations with precise pesticide application or assessing habitat quality in conservation settings (Ivosevic *et al.*, 2017). Pioneering efforts by Shields and Testa (1999) along with Kim *et al.* (2018) involved the utilization of a remote-controlled vehicle and a drone for insect sampling in the atmosphere over agricultural fields, showcasing the versatility of drone technology in entomological research. Currently, drones play a successful role in detecting, monitoring, and managing insect pest populations through precise pesticide application and assessing habitat quality in conservation settings (Ivosevic *et al.*, 2017).

### 13.7.1 Drone-mediated sampling

An apparatus attachable to drones is designed to trap airborne insects, functioning as either position fixed or freely movable. This innovative device incorporates double-charged dipolar electric field screen (DD-screen) coupled on drone, generating attractive force through an electric field to capture insects entering designated area. The strength of electric field is significant enough to prevent from escaping trap (Takikawa *et al.*,2020).

### 13.7.2 Types of drones used in sampling

The preference for rotary-wing drone designs over fixed-wing counterparts is largely attributed to their unique capabilities. These devices, commonly called as multicopters, harness continuous rotation of motors to create necessary airspeed for propellers to generate lift. Multicopters excel in precision control, allowing them to hover in place, take off, and land vertically within confined spaces, making them highly suitable for low-altitude operations like inspecting plant canopies for insects. On the other hand, fixed-wing drone designs are more advantageous at higher altitudes, particularly when the planetary boundary layer (PBL) is free of obstacles. This is due to their superior aerodynamic efficiency, enabling longer flight durations – an essential feature when targeting insects with low population densities. The fixed wing drones rely on advancing speed for lift generation, necessitating a continuous forward propulsion to create a consistent airflow over their wings, countering the force of gravity. Alternatively, they can maintain flight without active propeller rotation, as long as their relative airspeed remains persistent (Mulero-Pazmany *et al.*, 2022).

However, despite the promising potential of drone-based insect sampling, there are significant impediments to its widespread adoption. Legal restrictions prevent drones from operating beyond the operator's line of sight, during nighttime, and over restricted areas. These constraints limit the scope of drone operations. Nevertheless, there is a growing recognition that safe drone operations should be permitted in cases where they pose no threat to public safety. Many protected areas and agricultural regions already allow such operations.

Drones hold the transformative power to revolutionize insect aerial sampling, a method that can find applications in ecological research by assessing the severity of insect population declines. Furthermore, this technology has the potential to extend its reach into agriculture. Real-time data on insect populations obtained through drone-based sampling will contribute to the development of effective and sustainable crop supervision strategies.

## 13.8 DRONE-MEDIATED APPLICATION OF PESTICIDES

Actuation drones are deployed to manage pest hotspots, a more targeted approach compared to outdated aircraft pesticide spraying that often results in drift, causing ecological and health concerns (Al Heidary *et al.*, 2014). Drones equipped with spray technology offer advantages, including deployment ease, reduced operator exposure time, and possible spray drift saving (Faical *et al.*, 2014). Precision pesticide application, coupled with monitoring, has the potential to minimize overall sprays, reduce pesticide use, mitigate resistance development, and enhance natural enemies in the field, presenting a cost operative and ecologically sound solution. In realm of accuracy pest and disease management, RS technologies, particularly using drones, play a crucial role. ML algorithms prove valuable for accurate disease identification with extensive datasets (Ashourloo *et al.*, 2016). Various imaging techniques, including hyperspectral and multispectral, aid in distinguishing types of damage in crops (Zhang *et al.*, 2012). Hyperspectral technology, integrated with carbohydrate metabolic analysis, has demonstrated high precision in detecting Huanglongbing (HLB) across different seasons and cultivars (Weng *et al.*, 2018).

Recent advancements include the use of a RedEdge-MX camera on a UAV for multispectral imaging of a banana plantation infected with BFW (banana fusarium wilt) (Zhang *et al.*, 2022). Additionally, the early detection of the potato cyst nematode *Globodera pallida* before harvest is possible with UAV indicators, offering insights into infestation magnitude (Zhang *et al.*, 2022). In sugarcane fields, multispectral UAV photos combined with ML algorithms efficiently detect and segment white leaf disease (WLD), providing precise guidance for disease management (Narmilan *et al.*, 2022). These advancements underscore the potential for drone-based technologies in revolutionizing agricultural pest and disease management.

## 13.9 DRONE-MEDIATED PRECISION RELEASE OF BIOLOGICAL ENTITIES

Biological control offers a promising substitute to pesticides. This involves utilizing the population of one to manage the population of another harmful or undesirable organism. The agents employed include parasitoids, predators, fungi, bacteria, entomopathogenic nematodes, and viruses (Van Lenteren *et al.*, 2018). Biological pest control for agriculture entails parasites, and natural predators to reduce insect populations. As it targets certain pests while having a minimal influence on non-target species, this technique is seen to be a more environmentally friendly substitute for conventional chemical pesticides. One of the key advantages of biological pest control is that it targets only a single species of pests, protecting beneficial insects like pollinators from harm. This promotes ecological equilibrium and lessens the chance that pest populations will develop a tolerance to pesticides. The use of less pesticide overall on crops thanks to biological pest management can benefit both the environment and public health. Economic gains may also result from it since it can increase crop output while reducing pest management costs by protecting the population of natural enemies already present and avoiding the appearance of secondary pests. Farmers must monitor the populations of both pests and natural control agents to make sure that the balance is maintained. For instance, the discharge of natural predators or parasites must be timed properly to be most effective.

Beneficial insects released by drones may assist in overcoming application difficulties and promote greater use of biological pest management. Drones make it possible to release helpful insects more precisely and efficiently, saving money on labour and expanding coverage. Additionally, hiring a drone operator on short notice is much simpler than hiring a team of people to carry out a biocontrol application. Field study has demonstrated that the distribution and establishment of benefits from drone applications are of a high standard in addition to the increased efficiency. For augmentative biological control, actuation drones are useful to release huge natural entities for immediate pest control (Van Lenteren *et al.*, 2018). The advantage of using actuation drone tool is proper, and the localized distribution of natural enemies which increases the efficacy of biocontrol agent and also distribution cost is low compared to all other tools. Insect-killing fungi and nematodes, considered as natural enemies, are classically functional using conventional spray application apparatus (Piedra-Buena *et al.*, 2015). Drones can similarly be employed for the application of these biocontrol agents, akin to the application of pesticides through spray methods (Berner and Chojnacki, 2017).

Precision releases of natural enemies carried out by drones could be a highly valuable device for enhancive biological control, which relies on general deployment of natural enemies for quick management. They might allocate natural enemies precisely at targeted areas, which could upsurge the biocontrol agent's effectiveness and will decrease delivery costs. A predatory mite *Phytoseiulus persimilis Athias-Henriot* (Acari: *Phytoseiidae*) and parasitoid wasps like egg-parasitoid *Trichogramma* spp. (Hymenoptera: *Trichogrammatidae*) have drawn the majority of interest in the creation of drone-mounted dispensers. Dispensers mounted on motorcycles are a common practice among Brazilian farmers (Agronomic Nordeste 2015; Parra 2014), although this practice might harm crop. While the delivery of natural enemies by aircraft was suggested in 1980s (Herren *et al.*, 1987; Pickett *et al.*, 1987), the utilization of small drones presents numerous potentials. However, small drones would present a wide range of opportunities. more coverage than manual distribution, lower employment costs per acre, and maybe a greater reliance on natural enemies than on chemical sprays.

Moreover, employment of natural enemies is expensive and time taking. The natural enemy of two-spotted spider mite, *Phytoseiulus persimilis* Athias-Henriot (Acari: *Phytoseiidae*), is accessible in bottles with mineral substrate vermiculite, and the endorsed method of dispersion is through sprinkling plants (e.g., Koppert, 2017a; Biobest, 2019). Because of its extreme specialization, *P. persimilis* colonies perish in the absence of prey (Gerson and Weintraub, 2007; Dara, 2014). To aid in the spread of predators, a number of mechanical distribution methods have been created, including Mini-Airbug, a hand-held device along with a fan (Koppert, 2017b), and other gadgets (Opit *et al.*, 2005). France, Austria, Italy, Germany, and Canada are just a few of the countries that have started using drones for *Trichogramma* (Airborne Robotics, 2018). *Trichogramma* parasitoids are used in China to manage pests affecting sugarcane (*Saccharum* spp.). To control sugarcane borer (*Diatraea saccharalis*, Fabricius) in sugarcane, drones use the application of *Trichogramma* species and parasitoid *Cotesia flavipes* Cameron (Hymenoptera: Braconidae). Other lepidopteran pests of other crops are similarly combated with *Trichogramma* spp. (Xfly Brasil, 2017).

A start-up in California is providing facilities to disseminate predatory mites *via* drones on crops like strawberries to control two-spotted spider mite, a significant pest of many crops worldwide (Parabug, 2019). Another Australian business utilizes drones to spray strawberry crops with predatory mites (Drone Agriculture 2018). A drone-mounted tool is being created at the University of Queensland, Australia, to disperse predatory mites in maize (Pearl 2015). Collaborating with aeronautical engineering students from the University of California, Davis, Drs. Z. Kong and C. Nansen have developed a platform named BugBot for the drone-based dispersal of predatory mites (Teske *et al.*, 2002). Currently, they are researchers conducting tests on prototype and associated software to enhance the effectiveness of natural enemy release. Fostering cooperation among farmers, agricultural scientists, aerospace engineers, and software programmers is crucial for developing a product that is efficient and accessible.

Recently, ground and drone-mediated release of natural enemies is compared and assessed the effectiveness of *Trichogramma* species against *Choristoneura fumiferana*, Clemens (*Lepidoptera: Tortricidae*), a significant pest of fir and spruce trees in Canada and the United States. The ground releases by means of commercially available cards carrying parasitized eggs of Mediterranean flour moth *Ephestia kuehniella* Zeller (Lepidoptera: Pyralidae) were compared to drone-mediated releases using host eggs that had been parasitized by *Trichogramma* and mixed with vermiculite. According to the study, drone releases of natural enemies were quicker than terrestrial releases. These demonstrate the good efficient of drone-based *Trichogramma* dispersal in forests, particularly on small sizes and in situations where insecticide sprays are not suitable (Martel *et al.* 2018), even though additional research is required in this area. Similar research must be done in orchards and field crops to evaluate the effectiveness of natural enemies released by drones.

Other natural enemies will be employed using drones, including mealybug destroyer (*Cryptolaemus montrouzieri* Mulsant, Coleoptera: Coccinellidae), which controls mealybug, and green lacewing (*Chrysoperla* spp., Neuroptera: Chrysopidae) and minute pirate bug (*Oriusinsidiosus*, Hemiptera: Anthocoridae). This EcoDrone project involves EWH BioProduction, a company that produces beneficial organisms, along with Ecobotix, a business that provides drone-based services and is working on a different method of distributing natural enemies (Ecobotix 2018). For other varieties of helpful arthropods as well, drone-based dispensers might be modified.

The effectiveness of these procedures has received no peer-reviewed study so far. So, this is a plea for more investigation. Verifying that natural enemies dispersed via drones should be safe during transportation and delivery and efficient as biological control agents is of utmost significance. Additionally, the hardware and software systems that will accurately spread the natural enemies in various weather scenarios must be developed, especially in light of the fact that wind is a key component in the delivery. Distinct drone-mounted dispensers each employ a distinct technology that might be compared to optimize the dissemination of natural enemies. This might open the door for more extensive use of this resource.

## 13.10 DRONE-MEDIATED MATING DISRUPTION AND STERILE INSECT TECHNIQUE

One potential use of drones in pest control is delivery of sterile insects. In the New Zealand, Canada, and United States, drone-released sterile bug programmes were effective in monitoring codling moth numbers. Additionally, the Mexican fruit fly in citrus and the pink bollworm in cotton have both been successfully controlled in the United States by sterile insects launched by drones during experimental operations. It's possible to release sterile insects via drones. This will be less expensive and faster than terrestrial release, making sterile insect technique (SIT) a more widely used pest management technique. Therefore, another scope of drone use in pest management is sterile insect's release.

According to Judd and Gardiner (2005), codling moth (*Cydiapomonella* L.) is a significant issue in apple orchards (*Malus domestica* Borkh.). Pilot programmes using drones to disperse sterile insects have been effective in reducing codling moth populations in New Zealand, Canada, and the United States. Additionally, sterile insects released by drones proved successful in pilot programmes to control Mexican fruit fly *Anastrephaludens* Loew (Diptera: Tephritidae) and pink bollworm [*Pectinophoragossypiella* Saunders (Lepidoptera: Gelechiidae)] in citrus in the United States. In South African citrus orchards, the false codling moth [*Thaumatotibialeucotreta* Meyrick (Lepidoptera: Tortricidae)] could be effectively managed (FlyH2 Aerospace 2018).

Through irradiation, SIT creates completely or partly the sterile insects. Because there are also no progeny or sterile offspring after mating with wild insects, pest populations are decreased. SIT is a crucial IPM strategy since it is species-specific, ecologically benign, and well suited with other management techniques including biological control (Simmons *et al.* 2010). The sterile insect release via drone may be less expensive and quicker than land release, such as release by ATVs (all-terrain vehicles) or delivered by operated aircraft (Tan and Tan, 2013). Drone-dispersal could enhance moth performance in the case of sterile codling moths. ATVs release the moths on the orchard floor, while drones release them above the canopy. In order to help codling moths reach their favoured territory while reducing biotic and abiotic mortality causes, drone release may be beneficial. Codling moths like to mate in upper one third of canopy. Prior to orchard dispersion, irradiated moths must be transported in a refrigerator to prevent injury and scale loss. The capability of sterile moths to mate with wild ones may be improved with an improved transportation method from rearing capability to orchard (DuPont 2018). So, drone launches could increase the accessibility of SIT.

Additionally, mating disruptors like specialized pheromone and lure application technology (SPLAT) could be placed in marketable fields using drones (FlyH2 Aerospace 2018). SPLAT is a dollop-applicable inert matrix loaded with pheromones and/or insecticides (ISCA 2019a, b). In order to interrupt mating, pheromones must be released (Miller 2015), whereas attract followed by killing techniques use an attractant and a lethal agent (Gregg *et al.* 2018). In several cropping systems, including blueberry (*Vaccinium corymbosum* L.) and cranberry (Steffan *et al.* 2017), a combination of these techniques successfully manages a variety of pests. To enhance IPM practices in cranberry farms, University of Wisconsin researchers are presently creating a drone delivery device for SPLAT (Seely 2018).

### 13.10.1 Prevention of pest outbreak

Drones with sensing and actuation capabilities may help to stop pest outbreaks. Frequently, plants are too vulnerable to biotic and abiotic stresses like drought and nutritional deficits. This applies to a widespread range of arthropod pests, including spider mites (Ximénez-Embn *et al.* 2017), aphids (Lacoste *et al.* 2015), and lepidopteran larvae (Weldegergis *et al.* 2015). The precise application of abiotic stress release, like water and fertilizer, constitutes a significant strategy to mitigate risk of outbreaks by certain arthropod pests. This is based on the unshakable relationship amid abiotic stresses and the likelihood of arthropod pest outbreaks (West and Nansen 2014). In fact, the major focus of pest control may change from concentrating mostly on insecticide applications to more preventative strategy that places more emphasis on crop health (West and Nansen 2014). By determining plant stress levels and applying water and fertilizer in advance, sensing and actuation drone use may help bring about this transition. To the best of our knowledge, drones haven't been used for precise irrigation yet, and despite the fact that several drones on the market claim to be able to administer liquid or granular fertilizer, there isn't any peer-reviewed research on their application. In order to adequately respond to nutrient deficits discovered through remote sensing, several modern spray tractors offer choices for flexible rate treatments of nutrients (Raun *et al.*, 2002).

## 13.11 IoT IN UAVS

IoT has made advances in recent times and is often the most promising technology for driving agriculture. However, most farming regions lack base stations and Wi-Fi sites, preventing its use. But, the lack of wireless base stations and Wi-Fi stations in most rural regions is a major impediment to establishing surveillance systems. So, the data collected by the Wireless Sensor Network (WSN) cannot be sent wirelessly. An alternate method is to use UAVs to connect with WSNs over broad parts to get real-time data for giving out and investigation.

IoT and low-altitude RS technologies embodied by UAVs widely employed in monitoring spheres as information technology has advanced. IoT and UAV can monitor crop disease and pest incidence from ground micro and air macro lookouts, individually. By a network of affordable sensor nodes, IoT can gather real-time meteorological characteristics of crop growth. UAVs might assemble photographs of farms using spectral camera technology, which are used to analyse the incidence of agricultural pests and diseases. Deploying agricultural IoT devices on most farms, especially those in remote locations, presents challenges. To enhance energy-harvesting rates, a sun tracker device is designed to autonomously adjust an angle between solar panels and sunlight. A proposed flying mode aims to maximize wind force utilization and extend UAV flight duration. Photos captured by the UAV are subjected to spectrum analysis technology and then spread to a cloud data centre for assessment of damage caused by diseases and pests. In the agricultural context of China's Yangtze River Zone, the application of this framework reveals insights such as wheat sensitivity to disease occurring between temperatures of 14°C to 16°C and the inhibitory effect of heavy rainfall on the spread of wheat powdery mildew (Gao *et al.*, 2020).

### 13.11.1 Generating correct maps for farms

To properly detect diseases and pests in agriculture, data from precise farm maps acquired by drone must be analysed, to get the entire distribution of whole farm features based on picture analysis. The agricultural IoT stage enables an innovative and precise map-generating approach to make use of drone aerial photography and data given by farm sensors. The system specifically leverages the pictures created by drone cameras and values detected by sensors installed in soil to anticipate the potential of disease and pest emergence. The gateway of the system incorporates an ML employing probabilistic graphical models using Gaussian processes and hyperspectral data (Gao *et al.*, 2020).

### 13.11.2 Pests and diseases analysis through reflection spectrum

When crops are afflicted by pests and diseases, multiple factors like coverage, biomass, LAI (Leaf Area Index), moisture, leaf cell structure, nitrogen levels, pigment content, and overall appearance undergo changes. These modifications result in variations in reflectance spectrum, spanning from visible to thermal infrared range. Particularly noteworthy are significant differences in spectral properties between infrared and red regions compared to those of healthy ones. Monitoring reflectance spectrum allows for the determination of disease status of crops. The absorption of a substantial amount of energy by chlorophyll leads to distinct features in spectral reflectance of healthy crops, including a dip in red region (red valley) and a reduction in green zone, resulting in a pronounced green reflection zone (green peak). In terms of a quantifiable analysis, the relationship between chlorophyll content and spectral response was established in the 680–760 nm bands, focusing on red edge parameters such as the size and location of the red edge in the initial segment of the reflectance curve.

The spectrum shifts dramatically as the severity of disease increases. The red valley in red light array and green peak in green light array both progressively fade. The spectral reflectance of diseased wheat is substantially lesser than healthy wheat in NIR range. The damage caused by pests and diseases is tracked by using spectrum analysis, offers timely and accurate pest control information. Aerial images, depending on image processing technologies, assist to identify the existence of pests and diseases in crops (Gao *et al.*, 2020).

## 13.12 STANDARD OPERATING PROCEDURE ON THE USE OF DRONES FOR CROP PROTECTION COVERS

### 13.12.1 Pre-application

Avoid flying drones in drone-restricted areas like airports or electronic stations, but no prior authorization is needed in a green zone. The drone should be digital sky compliant and equipped with "No Permission–No Takeoff" hardware and firmware. Obtain a UIN (Unique Identification Number) and UAOP (Unmanned Aircraft Operator Permit) from DGCA in order to operate in regulated airspace, and attach it to drone. During flying, notice interference caused by mobile devices or signal blockage. Operators should be educated in drone operation and pesticide safety. Alcoholic beverages use should be prohibited within eight hours after operation. Standardize drone spray structure to guarantee proper nozzle output and labelled application rates. Generate smallest possible safeguard zones between drone-treated and non-target crops. Avoid spraying pesticides within 100 meters of water sources to prevent water pollution. Log your flights and notify the appropriate authorities of any issues or accidents. Without authorization, do not fly a drone over public places and events, or stadiums as well as over government amenities/military bases or over/near any no-fly drone zones. Except in regulated airspace near airports, seek prior authorization from private property without filing a flight plan or receiving AAI/ADC approval. Prohibit flying drone from a running vehicle, ship, or aircraft. Flying the drone in violation of the Unmanned Aircraft System Rules (UAS) and as updated from time to time is prohibited.

### 13.12.2 During application

Wear personal protective equipment (PPE) and read labels attentively throughout the application to understand safety guidelines. Eating, drinking, and smoking should be restricted throughout the application process. Confirm the flying path prior to application to minimize turn around. Spray with pure water for at least five minutes to test operation wherever practical and needed, and guarantee two-step dilutions to thoroughly dissolve pesticide to espouse suitable pressure for adjusted droplet spectrum (>100m). Weather elements such as wind speed, temperature, and appropriate humidity will all have an impact, so check them first. Ascertain an acceptable flying height over the target crop, as well as an adequate water volume and flying speed. Avoid going into treated areas, especially during vigorous bee foraging. Avoid spray drift on blossoming crops to protect humans, animals, and the environment. Strictly adhere to product label requirements and implement adequate risk-aversion procedures.

### 13.12.3 Post treatment

A prompt evacuation and relocation to fresh air is required. Hazardous waste should never be burned or buried in the usual course of business. Never leave empty containers in the field; properly dispose of them and keep them away from animals and water resources. Set up cautionary signs in the spray area to encourage individuals to be cautious. After each application, take a shower and change into clean clothes.

Several critical aspects must be examined, and good governance is required. There are various environmental constraints to drone use that require more research to overcome. Proper pilot training includes all of the necessary knowledge for the pilot. Drift management is critical, and non-targets must be protected first and foremost. Pesticide registration requirements for drone applications are required, along with the spray monitoring form and data reporting. The statutory provisions are critical and must be strictly observed.

## 13.13 CHALLENGES

### 13.13.1 Costs

The initial cost is high for drones along with different sensors and application techniques, increasing apparatus, and study software, which is a barrier to employ drones in precision crop protection. Although expenses are lowering as advances are made, they still remain expensive. Thus, a variety of firms are providing drone-related amenities including renting drones equipped with RS or providing predator dispersal services. Also, consulting firms provide RS and data analysis amenities, which may be coupled with other agriculture related amenities to give a single display place for effective record possession and scheduling.

### 13.13.2 Data collection, analysis, and interpretation

The reproducibility of RS data is a matter of concern with sensing drones. Canopy reflectance changes with sun angle, cloud coverage, and other variables.

As a result, comparing data collected on one day to data gained the subsequent day, or hour, is a challenging task. New ways for calibrating and analysing data from drones are constantly being developed, and this will be an ongoing process (Aasen *et al.*, 2018). Enhanced repetitions will make data more precise to enhance the precision in crop protection. Data analysis is another significant difficulty. Each hyperspectral sensor mission generates numerous data facing problems in storage, and analysis with various software with experts leads to significant time interval among data collection and results prominence. Multispectral data analysis is now significantly quicker than hyperspectral, but verdicts are not that accurate in the identification of pest and diseases (Yang *et al.*, 2009). Finally, data processing automation will immediately increase growers' use of precise hyperspectral datasets, resulting in faster identification and pest and disease hotspot discovery. Automated data processing will ease the connection between sensing and actuation drones, allowing to deliver solutions. Alternatively, one drone might serve as both a sensor and actuator, applying solutions directly to the desired target, which is still in early stages, with several obstacles to overcome. Information and experience from different domains must be merged so as to produce an actual actuation drone structure where agricultural scientists' expertise needed to reply scientific concerns followed by engineers and software designers to translate such information into hardware design followed by software components for effective and efficient solution distribution is a challenging one. Drones should be capable of autonomously computing and applying the best delivery plan. Flight and Payload flight duration and payload are main restraining variables in the use of drones. Developing a drone to identify insect hotspots and administer treatments is a challenging task. Indeed, the aforementioned AgriDrone can detect insect hotspots as well as deploy targeted remedies (OPTiM, 2016). To cover huge regions, however, communicating drones' network, or swarm, strengthens the efficiency (Gonzalez-de-Santos *et al.*, 2017). The sensing drones will identify pest hotspots and communicate with other actuation drones, precisely releasing biological control entities or agrochemicals as required. These drones can autonomously return to their base stations for recharging, eliminating the need for further human intervention. In drone community, establishing drone flights is an ongoing study field (Ponda *et al.*, 2015). Still, the discovery of a technique in such a prospective is not revealed and opens an era to work in such prospective.

### 13.13.3 Hostile weather circumstances and other environmental aspects

Unfavourable weather conditions pose obstacles to drone sensing and actuation events, and have an optimal operating temperature range. Robust winds can hinder the collection of airborne RS data along with the dissemination of pesticides and biocontrol agents. To minimize variations caused by uneven radiation scattering on natural surfaces, RS measurements are preferably collected in consistent sun and sensor angle geometry (Weyermann *et al.*, 2013). Collecting data at noon under bright, cloudless skies helps eliminate shadow effects and variations caused by shifting light intensity due to cloud cover (de Souza *et al.*, 2010). Though such ideal circumstances are not frequently encountered in farm settings. Drone flights are constrained by fog and clouds, and operating a drone in rain, hails, snow, or thunderstorms is not endorsed. Environmental variables like variations in height within fields or orchards and the presence of animals, particularly birds, further limit drone operations (Park *et al.*, 2012).

### 13.13.4 Guidelines and protocols

FAA has established Federal Aviation Regulations (FARs) to govern commercial and scientific use of drones in the United States. Before 2016, operating a drone required an expensive manned aircraft pilot license. Subsequently, a less restrictive remote pilot license for small drones became available, increasing accessibility (FAA 2016). Given that regulations are subject to frequent amendments, pilots must continually consider and adhere to all requirements. In the United States, drone pilots must maintain a visual line of sight (VLOS) at all times, limiting hovering to daylight hours, flying at or below 122 meters, and not exceeding 161 km/h. Flights over non-participating individuals are prohibited, and drones must yield to larger aircraft, including manned ones. The FAA can grant waivers for specific requirements, such as nighttime flying. Pre-flight inspections are mandatory for operational efficiency. Drones used for business or personal purposes in the United States must be registered with the FAA. Guidelines and registration procedures vary globally, so foreign users should adhere to country-specific regulations. In Brazil, where drones are widely used in precision agriculture, the National Agency of Civil Aviation (ANAC) imposed restrictions on civil and agricultural drone use in May 2017. As drone use becomes more common, standardized norms may emerge, integrating them seamlessly into routine operations (FAA, 2018; Stocker et al., 2017; Agencia Nacional de Aviacao Civil, 2017; Jorge et al., 2014).

### 13.13.5 Communication with growers

The widespread adoption of drones in profitable agricultural processes hinges on the willingness of producers to embrace technology. Growers will only accept expertise that has been proven to be effective, cost efficient, and compatible with existing practices. Achieving the highest-performing technology that aligns with the demands of producers necessitates widespread statement and teamwork among scientists, government institutes, industry experts, and commercial farmers (Lindblom *et al.*, 2017). Extension agents, dedicated to interpreting scientific results into practical, play a crucial role in fostering these connections through training and discourse.

### 13.13.6 Misuse

There is a potential for misuse that could violate people's privacy and involve the illegal transfer of information, posing a significant challenge for security purposes.

## 13.14 OPPORTUNITIES

Despite challenges, business opportunities for precision farming skills, including GPS, RS, GIS, and monitor systems, are vast in many developing countries. In India, Punjab and Haryana lead in farming mechanization and may be the first to implement these technologies on a large scale (Srinivasan, 1999). Several Asian countries, including Japan, have made unusual efforts to endorse precision farming, incorporating drones into the initiative (Shanwad *et al.*, 2004).

The application of drones in precision agriculture experienced a significant increase after 2017, attributed to reduced weight, lower costs, and increased payload ability. Drones utilized for crop health monitoring and recognition are predominantly of the multicopter and fixed wing types. The continuous reduction in size and price of drones, along with advancements in unmanned helicopters with high payload capacities, has become noteworthy. Drone cameras have evolved in terms of weight, size, and resolution, with a shift towards multispectral cameras to meet growing feature extraction requirements. The supervisors of drones have advanced basic microcontrollers to AI-enabled systems like Arduino Uno and Raspberry Pi. Drone technologies are transitioning from semi controlled to entirely automatic systems due to ongoing investigate. Integration of ML, AI, and IoT in drones has led to the development of farmer-friendly efficient systems. However, challenges persist in the adoption of drone technology in agriculture, including the cost, limited battery life, concerns about vision disruption, technology literacy among end users, and shortcomings in image processing and data analysis. Resolving these issues is crucial to increasing the adoption rate of drones in agriculture.

## 13.15 CONCLUSION

In the ever-evolving landscape of agriculture, the integration of drones into precision management of pests and diseases stands as a testament to the transformative power of technology. These aerial marvels have transcended conventional boundaries, ushering in a new era of innovation that promises to revolutionize how we safeguard crops, ecosystems, and food security.

With their capacity to swiftly traverse vast expanses of land and gather intricate data from above, drones have emerged as indispensable allies in the battle against pests and diseases. They offer a panoramic view that uncovers hidden stressors, tracks insidious invasions, and detects subtle anomalies that might otherwise elude human eyes. This unique vantage point empowers farmers, researchers, and environmentalists with insights that guide strategic decisions and targeted interventions, fostering a paradigm shift from broad-brush treatments to surgical precision. Drones have the potential to rewrite the script of sustainable agriculture. Their ability to precisely deploy treatments, reduce chemical usage, and optimize resource allocation not only amplifies productivity but also champions environmental stewardship. By mitigating the collateral damage often associated with traditional pest control methods, drones pave the way for a harmonious coexistence between agriculture and the ecosystems it depends upon.

Yet, this journey is not devoid of challenges. The technical complexities of drone operation, regulatory frameworks, data management intricacies, and the need for specialized expertise all form significant hurdles that demand careful navigation. However, these challenges stand as invitations to collaborate, innovate, and elevate the capabilities of both technology and human endeavour. In closing, drones in the precision management of pests and diseases represent a captivating chapter in the narrative of agricultural progress. Their wings carry not just the promise of healthier crops and more resilient ecosystems but also the potential to inspire a global community of farmers, scientists, and enthusiasts to reach new heights in the pursuit of sustainable and thriving food systems. As these aerial marvels continue to soar, they beckon us to explore uncharted horizons and cultivate a future where precision meets prosperity.

## REFERENCES

Aasen, H. and Bolten, A., 2018. Multi-temporal high-resolution imaging spectroscopy with hyperspectral 2D imagers–From theory to application. *Remote Sensing of Environment*, 205, pp. 374–389.

Adao, T., Peres, E., Padua, L., Hruska, J., Sousa, J.J. and Morais, R., 2017. UAS-based hyperspectral sensing methodology for continuous monitoring and early detection of vineyard anomalies. *Proceedings of the Small Unmanned Aerial Systems for Environmental Research*, Vila Real, Portugal, pp. 28–30.

Agencia Nacional De Aviacao Civil., Regas Da ANAC para uso De Dronesentramem Vigor, http://www.anac.gov.br/noticias/2017/, 2017.

Agronomic Nordeste, Trichobug (Trichogramma). Available from http://agromicnordeste.com.br/produtos, 2015.

Airborne Robotics, Agriculture & forestry. Available from https://www.air6systems.com/portfolio/agriculture-forestry, 2018.

Al Heidary, M., Douzals, J.P., Sinfort, C. and Vallet, A., 2014. Influence of spray characteristics on potential spray drift of field crop sprayers: A literature review. *Crop Protection*, 63, pp. 120–130.

Anthony, D., Elbaum, S., Lorenz, A. and Detweiler, C., 2014, September. On crop height estimation with UAVs. In *2014 IEEE/RSJ International Conference on Intelligent Robots and Systems* (pp. 4805–4812). IEEE.

Ashourloo, D., Aghighi, H., Matkan, A.A., Mobasheri, M.R. and Rad, A.M., 2016. An investigation into machine learning regression techniques for the leaf rust disease detection using hyperspectral measurement. *IEEE Journal of Selected Topics in Applied Earth Observations and Remote Sensing*, 9(9), pp. 4344–4351.

Bendig, J., Bolten, A. and Bareth, G., 2012. Introducing a low-cost mini-UAV for thermal-and multispectral-imaging. *The International Archives of the Photogrammetry, Remote Sensing and Spatial Information Sciences*, 39, pp. 345–349.

Berner, B. and Chojnacki, J. 2017. Influence of the air stream produced by the drone on the sedimentation of the sprayed liquid that contains entomopathogenic nematodes. *Journal of Research and Applications in Agricultural Engineering*, 62(3), pp. 26–29.

Berra, E. F., Gaulton, R. and Barr, S. (2017). Commercial off-the-shelf digital cameras on unmanned aerial vehicles for multitemporal monitoring of vegetation reflectance and NDVI. *IEEE Transactions on Geoscience and Remote Sensing*, 55(9), 4878–4886.

Biobest, Drones bundled with cameras or sensors. Available from https://www.blueskiesdronerental.com/product-category/rentals/drone-bundles, 2019.

Chinmay, B., Biswas, S.K. and Jat, M.L., 2007. Precision pest management: an emerging concept. *Ecofriendly Management of Plant Diseases*, pp. 105–110.

Chang, A., Jung, J., Maeda, M.M. and Landivar, J., 2017. Crop height monitoring with digital imagery from Unmanned Aerial System (UAS). *Computers and Electronics in Agriculture*, 141, pp. 232–237.

Costa, F.G., Ueyama, J., Braun, T., Pessin, G., Osório, F.S. and Vargas, P.A., 2012, July. The use of unmanned aerial vehicles and wireless sensor network in agricultural applications. *In 2012 IEEE International Geoscience and Remote Sensing Symposium* (pp. 5045–5048). IEEE.

Dara, S.K., 2014. Predatory mites for managing spider mites on strawberries. *UC ANR eJournal of Entomology and Biologicals.*

Drone Agriculture, Formerly aerobugs. Available from https://www.droneagriculture.com.au, 2018.

DuPont, T, Adding to the codling moth IPM tool box. *WSU Tree Fruit*. Available from http://treefruit.wsu.edu/article/adding-to-the-codling-moth-ipm-tool-box, 2018.

Ecobotix. Available from https://www.ecobotix.com/. In Danish, 2018.

Ehrlich, P.R. and Harte, J., 2015. To feed the world in 2050 will require a global revolution. *Proceedings of the National Academy of Sciences*, 112(48), pp. 14743–14744.

FAA, Press release – New FAA rules for small unmanned aircraft systems go into effect, https://www.faa.gov/news/press_releases/news_story.cfm?newsId=20734, 2016.

FAA, Unmanned Aircraft Systems getting started, https://www.faa.gov/uas/getting_started/, 2018.

Faical, B.S., Costa, F.G., Pessin, G., Ueyama, J., Freitas, H., Colombo, A., Fini, P.H., Villas, L., Osorio, F.S., Vargas, P.A. and Braun, T., 2014. The use of unmanned aerial vehicles and wireless sensor networks for spraying pesticides. *Journal of Systems Architecture*, 60(4), pp. 393–404.

FlyH2 Aerospace, Agriculture - greenfly aviation. Available from https://flyh2.com/agriculture-greenfly-aviation/, 2018.

Gao, D., Sun, Q., Hu, B. and Zhang, S., 2020. A framework for agricultural pest and disease monitoring based on internet-of-things and unmanned aerial vehicles. *Sensors*, 20(5), p.1487.

Gerson, U. and Weintraub, P.G., 2007. Mites for the control of pests in protected cultivation. *Pest Management Science: Formerly Pesticide Science*, 63(7), pp. 658–676.

Giles, D. and Billing, R., 2015. Deployment and performance of a UAV for crop spraying. *Chemical Engineering Transactions*, 44, pp. 307–312.

Gonzalez-de-Santos, P., Ribeiro, A., Fernandez-Quintanilla, C., Lopez-Granados, F., Brandstoetter, M., Tomic, S., Pedrazzi, S., Peruzzi, A., Pajares, G., Kaplanis, G. and Perez-Ruiz, M., 2017. Fleets of robots for environmentally-safe pest control in agriculture. *Precision Agriculture*, 18, pp. 574–614.

Gregg, P.C., Del Socorro, A.P. and Landolt, P.J., 2018. Advances in attract-and-kill for agricultural pests: beyond pheromones. *Annual Review of Entomology*, 63, pp. 453–470.

Herren, H., Bird, T.J. and Nadel, D.J., 1987. Technology for automated aerial release of natural enemies of the cassava mealybug and cassava green mite. *International Journal of Tropical Insect Science*, 8, pp. 883–885.

Hobbs, J.E., 2020. Food supply chains during the COVID-19 pandemic. *Canadian Journal of Agricultural Economics/Revue Canadienne D'agroeconomie*, 68(2), pp. 171–176.

Huang, J., Wang, X., Li, X., Tian, H. and Pan, Z., 2013. Remotely sensed rice yield prediction using multi-temporal NDVI data derived from NOAA's-AVHRR. *PLoS One*, 8(8), p.e70816.

Huang, Y., Hoffmann, W.C., Lan, Y., Wu, W. and Fritz, B.K., 2009. Development of a spray system for an unmanned aerial vehicle platform. *Applied Engineering in Agriculture*, 25(6), pp. 803–809.

ISCA, Attract & kill: the hybrid IPM solution. Available from https://www.iscatech.com/solutions/attract-kill/, 2019b.

ISCA, Mating disruption. Available from https://www.iscatech.com/solutions/mating-disruption/, 2019a.

Ivosevic, B., Han, Y.G. and Kwon, O., 2017. Monitoring butterflies with an unmanned aerial vehicle: current possibilities and future potentials. *Journal of Ecology and Environment*, 41, pp. 1–6.

Jorge, L.A., Brandão, Z.N. and Inamasu, R.Y., 2014, October. Insights and recommendations of use of UAV platforms in precision agriculture in Brazil. In *Remote Sensing for Agriculture, Ecosystems, and Hydrology XVI* (Vol. 9239, pp. 313–330). SPIE.

Judd, G.J. and Gardiner, M.G., 2005. Towards eradication of codling moth in British Columbia by complimentary actions of mating disruption, tree banding and sterile insect technique: five-year study in organic orchards. *Crop Protection*, 24(8), pp. 718–733.

Kale, S.D., Khandagale, S.V., Gaikwad, S.S., Narve, S.S. and Gangal, P.V., 2015. Agriculture drone for spraying fertilizer and pesticides. *International Journal of Advanced Research in Computer Science and Software Engineering*, 5(12), pp. 804–807.

Keller, M.D. and Shields, E.J., 2014. Aerobiological sampling efficiency of media-containing Petri plates for use in lower atmosphere spore collection. *Aerobiologia*, 30, pp. 103–109.

Kim, H.G., Park, J.S. and Lee, D.H., 2018. Potential of unmanned aerial sampling for monitoring insect populations in rice fields. *Florida Entomologist*, 101(2), pp. 330–334.

Koppert, Mini-airbug. Available from https://www.koppert.com/products/distribution-appliances/mini-airbug/, 2017b.

Koppert, Spidex - Phytoseiulus persimilis. Available from https://www.koppert.com/products/products-pests-diseases/spidex/, 2017a.

Lacoste, C., Nansen, C., Thompson, S., Moir-Barnetson, L., Mian, A., McNee, M. and Flower, K.C., 2015. Increased susceptibility to aphids of flowering wheat plants exposed to low temperatures. *Environmental Entomology*, 44(3), pp. 610–618.

Lindblom, J., Lundström, C., Ljung, M. and Jonsson, A., 2017. Promoting sustainable intensification in precision agriculture: review of decision support systems development and strategies. *Precision Agriculture*, 18, pp. 309–331.

Martel, V., Trudeau, S., Johns, R., Owens, E., Smith, S.M. and Bovin, G., 2018. *Testing the efficacy of Trichogramma minutum in the context of an 'Early Intervention Strategy'against the spruce budworm using different release methods.* SERG-i Annual Reports. Quebec, Canada, pp. 276–283.

Miller, N., 2015. *CALS researchers deploy insect 'birth control'to protect cranberries.* University of Wisconsin-Madison News.

Moran, M.S., Inoue, Y. and Barnes, E.M., 1997. Opportunities and limitations for image-based remote sensing in precision crop management. *Remote Sensing of Environment*, 61(3), pp. 319–346.

Mulero-Pazmány, M., Martinez-de Dios, J.R., Popa-Lisseanu, A.G., Gray, R.J., Alarcon, F., Sanchez-Bedoya, C.A., Viguria, A., Ibanez, C., Negro, J.J., Ollero, A. and Marron, P.J., 2022. Development of a fixed-wing drone system for aerial insect sampling. *Drones*, 6(8), p.189.

Narmilan, A., Gonzalez, F., Salgadoe, A.S.A. and Powell, K., 2022. Detection of white leaf disease in sugarcane using machine learning techniques over UAV multispectral images. *Drones*, 6(9), p.230.

Nijland, W., De Jong, R., De Jong, S.M., Wulder, M.A., Bater, C.W. and Coops, N.C., 2014. Monitoring plant condition and phenology using infrared sensitive consumer grade digital cameras. *Agricultural and Forest Meteorology*, 184, pp. 98–106.

Opit, G.P., Nechols, J.R., Margolies, D.C. and Williams, K.A., 2005. Survival, horizontal distribution, and economics of releasing predatory mites (Acari: Phytoseiidae) using mechanical blowers. *Biological Control*, 33(3), pp. 344–351.

OPTiM, OPTiM's AgriDrone undergoes the world's first successful trials for insect extermination by drone. Available from https://en.optim.co.jp/news-detail/11172, 2016.

Parabug, Parabug, biocontrol by drone. Available from https://www.parabug.solutions/, 2019.

Park, C.Y., Jang, B.W., Kim, J.H., Kim, C.G. and Jun, S.M., 2012. Bird strike event monitoring in a composite UAV wing using high speed optical fiber sensing system. *Composites Science and Technology*, 72(4), pp. 498–505.

Parra, J.R.P., 2014. Biological control in Brazil: An overview. *Scientia Agricola*, 71, pp. 420–429.

Pearl, E. *Drone used to drop beneficial bugs on corn crop*. The University of Queensland, Australia, News (UQ News). Available from https://www.uq.edu.au/news/article/2015/04/drone-used-drop-beneficial-bugs-corn-crop, 2015.

Pickett, C.H., Gilstrap, F.E., Morrison, R.K. and Bouse, L.F., 1987. Release of predatory mites (Acari: Phytoseiidae) by aircraft for the biological control of spider mites (Acari: Tetranychidae) infesting corn. *Journal of Economic Entomology*, 80(4), pp. 906–910.

Ponda, S.S., Johnson, L.B., Geramifard, A. and How, J.P., 2015. Cooperative mission planning for multi-UAV teams. *Handbook of Unmanned Aerial Vehicles*, 2, pp. 1447–1490.

Primicerio, J., Di Gennaro, S.F., Fiorillo, E., Genesio, L., Lugato, E., Matese, A. and Vaccari, F.P., 2012. A flexible unmanned aerial vehicle for precision agriculture. *Precision Agriculture*, 13(4), pp. 517–523.

Radoglou-Grammatikis, P., Sarigiannidis, P., Lagkas, T. and Moscholios, I., 2020. A compilation of UAV applications for precision agriculture. *Computer Networks*, 172, p.107148.

Raun, W.R., Solie, J.B., Johnson, G.V., Stone, M.L., Mullen, R.W., Freeman, K.W., Thomason, W.E. and Lukina, E.V., 2002. Improving nitrogen use efficiency in cereal grain production with optical sensing and variable rate application. *Agronomy Journal*, 94(4), pp. 815–820.

Mrema, G., Soni, P. and Rolle, R.S., 2014. *A regional strategy for sustainable agricultural mechanization: sustainable mechanization across agri-food chains in Asia and the Pacific region*. RAP Publication, (2014/24).

Roth, L. and Streit, B., 2018. Predicting cover crop biomass by lightweight UAS-based RGB and NIR photography: an applied photogrammetric approach. *Precision Agriculture*, 19, pp. 93–114.

Salami, E., Barrado, C. and Pastor, E., 2014. UAV flight experiments applied to the remote sensing of vegetated areas. *Remote Sensing*, 6(11), pp. 11051–11081.

Samways, M.J., McGeoch, M.A. and New, T.R., 2010. *Insect conservation: a handbook of approaches and methods*. Oxford University Press.

Sarghini, F. and De Vivo, A., 2017a. Interference analysis of an heavy lift multirotor drone flow field and transported spraying system. *Chemical Engineering Transactions*, 58, pp. 631–636.

Sarghini, F. and De Vivo, A., 2017b. Analysis of preliminary design requirements of a heavy lift multirotor drone for agricultural use. *Chemical Engineering Transactions*, 58, pp. 625–630.

Seely, R., Drones, joysticks, and data-driven farming, In Grow. *University of Wisconsin-Madison College of Agricultural and LifeSciences*. Available from https://grow.cals.wisc.edu/wp-content/uploads/sites/14/2018/06/Grow-Summer2018-web.pdf, pp. 16–21, 2018.

Shanwad, U.K., Patil, V.C. and Gowda, H.H., 2004. *Precision farming: dreams and realities for Indian agriculture*. Map India.

Piedra-Buena, A., López-Cepero, J. and Campos-Herrera, R., 2015. Entomopathogenic nematode production and application: regulation, ecological impact and non–target effects. In *Nematode Pathogenesis of Insects and Other Pests: Ecology and Applied Technologies for Sustainable Plant and Crop Protection* (pp. 255–282). Cham: Springer International Publishing.

Shields, E.J. and Testa, A.M., 1999. Fall migratory flight initiation of the potato leafhopper, Empoasca fabae (Homoptera: Cicadellidae): Observations in the lower atmosphere using remote piloted vehicles. *Agricultural and Forest Meteorology*, 97(4), pp. 317–330.

Simmons, G.S., Suckling, D.M., Carpenter, J.E., Addison, M.F., Dyck, V.A. and Vreysen, M.J.B., 2010. Improved quality management to enhance the efficacy of the sterile insect technique for lepidopteran pests. *Journal of Applied Entomology*, 134(3), pp. 261–273.

de Souza, E.G., Scharf, P.C. and Sudduth, K.A., 2010. Sun position and cloud effects on reflectance and vegetation indices of corn. *Agronomy Journal*, 102(2), pp. 734–744.

Srinivasan, A., 1999, January. Precision farming in Asia: progress and prospects. In *Proceedings of the Fourth International Conference on Precision Agriculture* (pp. 623–639). Madison, WI: American Society of Agronomy, Crop Science Society of America, Soil Science Society of America.

Steffan, S.A., Chasen, E.M., Deutsch, A.E. and Mafra-Neto, A., 2017. Multi-species mating disruption in cranberries (Ericales: Ericaceae): Early evidence using a flowable emulsion. *Journal of Insect Science*, 17(2), p.54.

Stephens, E.C., Martin, G., Van Wijk, M., Timsina, J. and Snow, V., 2020. Impacts of COVID-19 on agricultural and food systems worldwide and on progress to the sustainable development goals. *Agricultural Systems*, 183, p.102873.

Stocker, C., Bennett, R., Nex, F., Gerke, M. and Zevenbergen, J., 2017. Review of the current state of UAV regulations. *Remote Sensing*, 9(5), p.459.

Surovy, P., Almeida Ribeiro, N. and Panagiotidis, D., 2018. Estimation of positions and heights from UAV-sensed imagery in tree plantations in agrosilvopastoral systems. *International Journal of Remote Sensing*, 39(14), pp. 4786–4800.

Takikawa, Y., Matsuda, Y., Nonomura, T., Kakutani, K., Kusakari, S.I. and Toyoda, H., 2020. Exclusion of flying insect pests from a plastic hoop greenhouse by a bamboo blind-type electric field screen. *Journal of Agricultural Science*, 12(2), p.50.

Tan, L.T. and Tan, K.H., 2013. Alternative air vehicles for sterile insect technique aerial release. *Journal of Applied Entomology*, 137, pp. 126–141.

Teske, M.E., Bird, S.L., Esterly, D.M., Curbishley, T.B., Ray, S.L. and Perry, S.G., 2002. AgDrift®: A model for estimating near-field spray drift from aerial applications. *Environmental Toxicology and Chemistry: An International Journal*, 21(3), pp. 659–671.

Torres-Sanchez, J., de Castro, A.I., Pena, J.M., Jimenez-Brenes, F.M., Arquero, O., Lovera, M. and Lopez-Granados, F., 2018. Mapping the 3D structure of almond trees using UAV acquired photogrammetric point clouds and object-based image analysis. *Biosystems Engineering*, 176, pp. 172–184.

van Lenteren, J.C., Bolckmans, K., Köhl, J., Ravensberg, W.J. and Urbaneja, A., 2018. Biological control using invertebrates and microorganisms: plenty of new opportunities. *BioControl*, 63, pp. 39–59.

Varshney, D., Roy, D. and Meenakshi, J.V., 2020. Impact of COVID-19 on agricultural markets: assessing the roles of commodity characteristics, disease caseload and market reforms. *Indian Economic Review*, 55, pp. 83–103.

Weldegergis, B.T., Zhu, F., Poelman, E.H. and Dicke, M., 2015. Drought stress affects plant metabolites and herbivore preference but not host location by its parasitoids. *Oecologia*, 177, pp. 701–713.

Weng, H., Lv, J., Cen, H., He, M., Zeng, Y., Hua, S., Li, H., Meng, Y., Fang, H. and He, Y., 2018. Hyperspectral reflectance imaging combined with carbohydrate metabolism analysis for diagnosis of citrus Huanglongbing in different seasons and cultivars. *Sensors and Actuators B: Chemical*, 275, pp. 50–60.

West, K. and Nansen, C., 2014. Smart-use of fertilizers to manage spider mites (Acari: Tetrachynidae) and other arthropod pests. *Plant Science Today*, 1(3), pp. 161–164.

Weyermann, J., Damm, A., Kneubühler, M. and Schaepman, M.E., 2013. Correction of reflectance anisotropy effects of vegetation on airborne spectroscopy data and derived products. *IEEE Transactions on Geoscience and Remote Sensing*, 52(1), pp. 616–627.

Xfly Brasil, Lancador de TrichogrammaGranel. Available from https://www.xflybrasil.com/trichogramma, 2017.

Ximenez-Embun, M.G., Castanera, P. and Ortego, F., 2017. Drought stress in tomato increases the performance of adapted and non-adapted strains of Tetranychus urticae. *Journal of Insect Physiology*, 96, pp. 73–81.

Yang, C., Everitt, J.H., Bradford, J.M. and Murden, D.A.L.E., 2009. Comparison of airborne multispectral and hyperspectral imagery for estimating grain sorghum yield. *Transactions of the ASABE*, 52(2), pp. 641–649.

Yao, L., Jiang, Y., Zhiyao, Z., Shuaishuai, Y. and Quan, Q., 2016, August. A pesticide spraying mission assignment performed by multi-quadcopters and its simulation platform establishment. In *2016 IEEE Chinese Guidance, Navigation and Control Conference (CGNCC)* (pp. 1980–1985). IEEE.

Zhang, D., Zhou, X., Zhang, J., Lan, Y., Xu, C. and Liang, D., 2018. Detection of rice sheath blight using an unmanned aerial system with high-resolution color and multispectral imaging. *PloS One*, 13(5), p.e0187470.

Zhang, S., Li, X., Ba, Y., Lyu, X., Zhang, M. and Li, M., 2022. Banana Fusarium Wilt disease detection by supervised and unsupervised methods from UAV-based multispectral imagery. *Remote Sensing*, 14(5), p.1231.

Zhang, J. C., Pu, R. L., Wang, J. H., Huang, W. J., Yuan, L. and Luo, J. H. (2012). Detecting powdery mildew of winter wheat using leaf level hyperspectral measurements. *Computers and Electronics in Agriculture*, 85, 13–23.

Zhu, H., Lan, Y., Wu, W., Hoffmann, W.C., Huang, Y., Xue, X., Liang, J. and Fritz, B., 2010. Development of a PWM precision spraying controller for unmanned aerial vehicles. *Journal of Bionic Engineering*, 7(3), pp. 276–283.

# Index

Pages in **bold** refer to tables.